ANNA BARGAGLIOTTI & CHRISTINE FRANKLIN

Statistics and Data Science for Teachers

Library of Congress Cataloging-in-Publication Data

Names: Bargagliotti, Anna, author. | Franklin, Christine, author.

Title: Statistics and data science for teachers / Anna Bargagliotti, Christine Franklin.

Description: Alexandria : American Statistical Association, 2021. | Summary: "A main goal of Statistics and Data Science for Teachers is to provide teacher educators with a resource to guide entire courses and professional development, or portions of courses and professional development when preparing teachers of all school level grade levels to teach the foundations of statistics and data science in their classrooms. In supporting the spirit of Pre-K-12 Guidelines for Assessment and Instruction in Statistics Education II (GAISE II), this book presents statistical ideas through investigations and engagement with the statistical problem-solving process of formulating statistical investigative questions, collecting/considering data, analyzing data, and interpreting results"— Provided by publisher.

Identifiers: LCCN 2021048141 | ISBN 9781734223545 (trade paperback)

Subjects: LCSH: Statistics — Study and teaching. | Statistics — Data processing. | AMS: Statistics — Instructional exposition (textbooks, tutorial papers, etc.) pertaining to statistics. | Mathematics education—Education of combinatorics, graph theory, probability theory, and statistics — Descriptive statistics (educational aspects). | Mathematics education — Education of combinatorics, graph theory, probability theory, and statistics — Foundations and methodology of statistics (educational aspects).

Classification: LCC QA276.18 .B37 2021 | DDC 519.5071—dc23/eng/20211108

LC record available at https://lccn.loc.gov/2021048141

Printed in the United States of America

10 9 8 7 6 5 4 3 2 1

978-1-7342235-4-5

Table of Contents

outlines how to achieve this goal. To reach the goal of a data-literate population, teachers must be prepared to deliver statistics and data-science content in the classroom. The importance of teacher preparation in statistics and data science has been further articulated in the NCTM document *Catalyzing Change in High School Mathematics: Initiating Critical Conversations,* wherein statistics is included as a major content area along with algebra and geometry. In addition, as data science becomes more prominent in state standards across the nation, teachers need preparation to meet these demands.

Importance of Data Today. As noted in *GAISE II,* in recent years, there has been an increased emphasis on data and data-driven decisionmaking in our society. Data, both traditional types and nontraditional types, are everywhere, and statistics is often called the science of data. The collection of data, the quantification of our lives, and the reliance on data for decisionmaking are prominent features in today's data-driven society. For example, consider the recent popularity of body-tracking devices that quantify our movements, diets, and other daily behaviors. Devices such as Fitbits, Apple Watches, and cellular phones allow each individual to collect data about themselves on a daily basis. Gould's article "Statistics and the Modern Student" (2011) provides a wealth of examples of data being collected by various technologies throughout a typical day. For example, he mentions the Pandora app making music suggestions for his listening pleasure based on his prior song choices and his responses to preference questions asked by Pandora. Another example Gould provides is related to geomaps updating gas prices in the greater Los Angeles area so that an individual can make an informed decision on where to purchase gas.

These examples demonstrate that we can collect, analyze, and interpret data to answer statistical investigative questions and to drive decisions in our lives. Presumably we could then analyze and interpret the data to help us make life choices, such as how many steps we should take per day, which foods we should eat, or which music suggestions we should follow.

Because increasing importance is being placed on data literacy throughout society—for example, statistical information flooding the news—data increasingly influence our lives. This has dramatically been demonstrated since March 2020, when the world entered the COVID-19 pandemic. As engaged members of society, we must be educated on how to make sense of statistical information. The demand for people educated in statistics and data science has grown. For example, jobs related to statistics are expected to grow by about 27% between 2012 and 2022, according to the Bureau of Labor Statistics (2013).

The State of Statistics Education. As data become increasingly more accessible and the action of collecting data continues to be present in our daily lives, it is necessary to prepare a statistically proficient population. Because of this new emphasis, data and statistics have become a key component of K–12 mathematics curricula across the country. The current standards contain a substantial amount of statistics at the middle- and high-school levels. As states revise their math standards, additional statistics standards are being added, particularly at the elementary level. In addition, the number of students taking Advanced Placement (AP) statistics in high school has drastically increased over the past several years—from 7,500 in 1997 to more than 222,000 in 2019, illustrating the student demand for statistics education (https://secure-media.collegeboard.org/digitalServices/pdf/research/2019/Program-Summary-Report-2019.pdf). Given these increased demands, it is imperative that teachers be prepared accordingly.

While data in our daily lives have become increasingly more important, the idea of promoting statistics in schools is not new. In 1923, the Mathematical Association of America's (MAA's) report *The Reorganization of Mathematics in Secondary Schools* included two recommendations. First, statistics should be offered in middle-school mathematics; second, statistics should be included as a required course in high school. Today, almost 100 years later, the Common Core State Standards in Mathematics (CCSSM) and other state standards include statistics as part of the mathematics curriculum in schools. Unfortunately, statistics still struggles to have a place in teacher-preparation programs and professional development. This book was written with the intention of addressing this issue. A goal of this book is to provide a resource to guide teacher preparation in statistics and data science.

Teacher Preparation. Teachers need to understand the statistics they must teach according to current state standards and be familiar with appropriate methods and technologies for teaching statistics and data science. Unfortunately, teachers have been given little opportunity to learn statistics in their training programs (*MET II*, 2012). Currently, there are resources available on the ASA website, at www.amstat.org/AMSTAT/Education/K-12-Educators/ASA/Education/K-12-Educators.aspx?hkey=66767c1e-fea8-43ea-8665-12bf718997f6, to use in teachers' preparation; there are also online professional-development courses developed by Dr. HollyLynne S. Lee at www.mooc-list.com/instructor/hollylynne-s-lee. Other resources for teaching statistics and data science also exist; however, a comprehensive book, not focused on AP statistics, that covers statistics and data-science preparation for teachers is currently lacking. In fact, the *MET II* report has identified statistics as the one area in which teachers have the largest need in both content and pedagogy (CBMS, 2011). Despite numerous approaches to teacher education in general mathematics (e.g., Ball, 1991; Ball & Bass, 2000; Hill & Ball, 2004; Franke et al., 2009), there is a shortage of comprehensive resources for training teachers to teach statistics and data science.

Purpose of This Book. Access to resources that addresses current school-level standards and recommendations put forth in the *SET* report would empower teachers and teacher educators to teach statistics and data science in a way that is rich and relevant. **This book aims to provide teachers with a foundation in statistics and data as outlined by *SET* and included in state standards.** In the spirit of *GAISE II,* this book presents statistical ideas through investigations and engagement with the statistical problem-solving process of formulating statistical investigative questions, collecting/considering data, analyzing data, and interpreting results. For each investigation, worksheets prepared by teachers to be used in the classroom can be downloaded here https://bit.ly/Statistics-DataScience-for-Teachers.

This book, *Statistics and Data Science for Teachers,* encompasses all grade bands of teacher preparation (elementary, middle, and high) up to the content of an AP statistics course. The authors envision that it could be used to guide entire courses and professional development, or portions of courses and professional development that teachers may be taking. **A main goal of the book is to provide teacher educators with a resource to use when preparing teachers of all grade levels to teach statistics and data science in their classrooms.**

The material presented in this book has been tested and used in numerous teacher-preparation settings, such as preservice teacher preparation, graduate courses in statistics specifically for in-service teachers, and professional developments for districts, in both face-to-face and online modalities. In its entirety, the material in this book takes approximately 50 hours to address in a face-to-face class. Each investigation provided takes approximately one hour to explore, and additional materials, such as extra practice problems, make up an additional 15 hours. The book can be presented in multiple ways. The investigations could be adapted for use in an online forum or a distance-learning opportunity. Additionally, the book is structured in a way that an individual teacher could read the book and self-instruct by following along with each investigation. When reading through the book, one will find that the materials build on one another. Thus someone who is self-directed can follow the progression of the topics and discussion to develop a foundation in statistics. The book presents material in a conversational manner, not a formal textbook style, thus making it easy and engaging to read.

The goal of this book is to provide guidance in preparing educators in a way that helps teachers gain:

- an understanding of statistics and data-science content covered in grades K–12,
- an appreciation of and a familiarity with using technology such as apps and statistical software, and
- an understanding of how to think about data.

Structure of This Book. Throughout this book, teachers are viewed as active participants, working to deepen their understanding of modern-day statistics. The book presents content in the form of case studies and investigations. Case studies use news articles to motivate the statistics content. The investigations introduce content in the spirit of the *Pre-K–12 GAISE* guidelines, wherein each investigation is guided by a question, and the scope of the investigation is to answer the question through data. Throughout the book, we make reference to "Mathematical Practices through a Statistical Lens," described in Chapter 3 of the *SET* report. These practices describe the habits of mind one might employ while problem-solving in the statistics content domain. As noted in the *SET* report, these practices differ in important ways from the mathematical practices. In statistics, the practices focus on sifting through uncertainty and variability in systematic ways.

The book is organized into three sections:

- "Statistics as a Problem-Solving Process"
- "Toward Data Science"
- "Probability Unpacked"

Within each of these sections, standard topics of descriptive statistics, associations and relationships, distributions, probability, and sampling distributions are all developed. In addition, the "Toward Data Science" section illustrates how the principles of data science can be delivered in K–12 throughout all of the grade bands.

We are excited about sharing the insights that we have learned from our experiences as teachers and from collaborating with amazing teachers and students. It is our hope that you find these materials creative, realistic, and helpful for inspiring the development of sound statistical-reasoning skills.

Anna Bargagliotti & Christine Franklin

UNIT 1: Statistics as a Problem-Solving Process

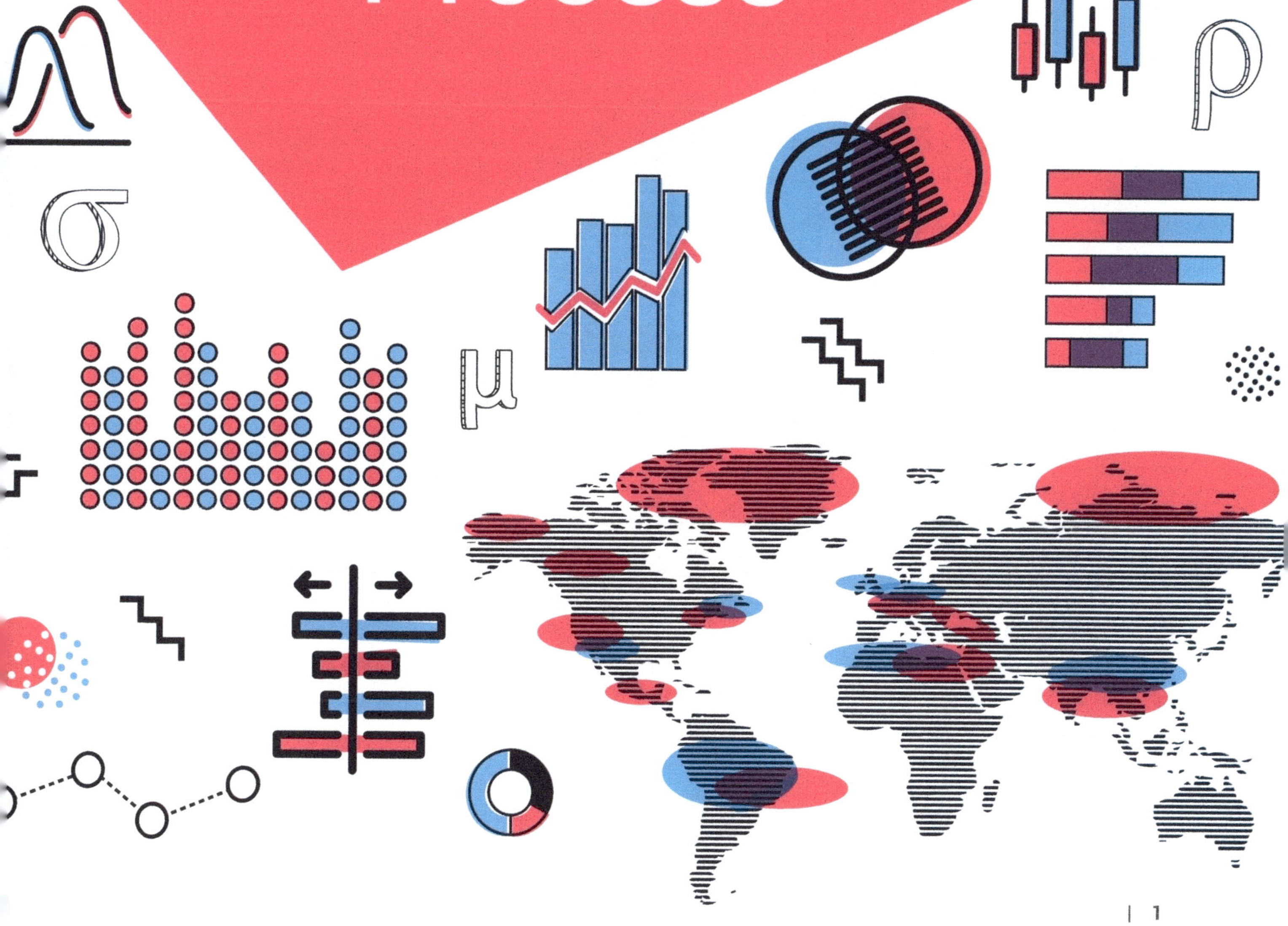

UNIT 1A: The Statistical Problem-Solving Process

In 2020, the *Pre-K–12 Guidelines for Assessment and Instruction in Statistics Education II (GAISE II): A Framework for Statistics and Data Science Education* report articulated the statistical thinking process. The GAISE II framework described statistics as a process (illustrated below) that includes four components: Formulate Statistical Investigative Questions, Collect/Consider the Data, Analyze the Data, and Interpret the Results (*GAISE II*, 13).

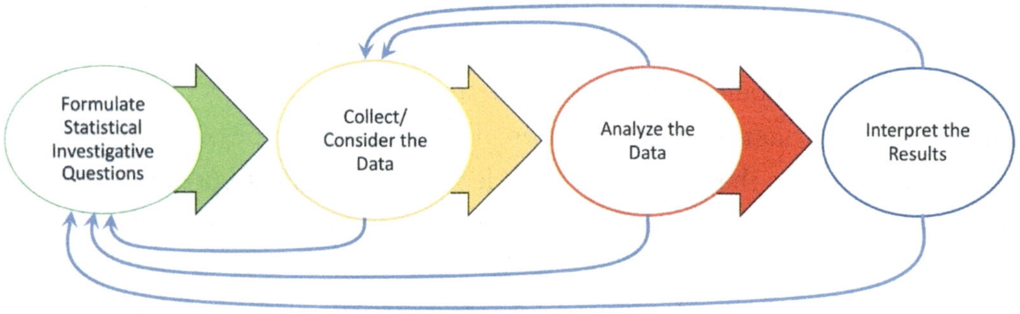

Investigative questions provide a starting point for statistical investigations to be carried out. They can be thought of as research questions. Such questions necessitate data to be collected in order for them to be answered. Such questions anticipate variability in data and aim to understand the variability. Data are collected in such a way that when analyzed, they will provide a pathway to answer the investigative questions; analyses are conducted to summarize the data in order to better understand the variability present in the data and to answer the investigative question; and interpretations are drawn that provide an answer to the investigative question posed. Throughout this book, we will refer to the process of going through the four phases in Figure 1 as the **statistical problem-solving process.**

GAISE II explicitly discussed how questioning plays an important role throughout *all* components of the process. In fact, in statistics, there are several different layers of questioning within the statistical problem-solving process (Arnold and Franklin, 2021). For example, there are questions that motivate the study investigation, statistical questions that motivate the data collection, questions that produce data (e.g., survey questions), questions that prompt analyses of the data (e.g. What are typical values?), interrogative questions to understand the data (e.g., What is the unit of observation?), and questions that focus on the interpretation of results. For each of the components in Figure 1, questioning plays an important role. We will discuss the role of questions in statistics in more depth in the following unit. For now, the important thing to note is that the collection, analysis, and interpretation of data are all motivated by the necessity to answer the investigative question posed.

Investigation 1A.1: Third Grade Sports

Goals for this investigation: Illustrate the statistical investigative process—formulate statistical investigative questions, collect/consider the data, analyze the data, and interpret the results.

At a local elementary school, school administrators are interested in purchasing new equipment for the third graders' recess time. They do not know what to buy, so they ask: What type of equipment should be purchased for the playground? This question motivates the administrators to conduct an investigation. To decide what type of equipment they should purchase, they realize they need to first understand the students' preferences on what they would use. The administrators pose the following investigative question:

What sports do third grade students at our school prefer?[1]

This question is an investigative question because it is unlikely that all the third grade students will have the same preference, and thus data on preferences will vary. To answer this investigative question, they decide to survey the entire third grade class by asking them this survey question, which will produce data to analyze:

What is your favorite sport to play at school?

Basketball — *Soccer*
Football — *Tag*
Handball

1 "What sports do third grade students at our school prefer?" is the statistical question that motivates the data collection. We will refer to questions used to motivate data collection as "investigative questions" throughout the book. Such questions initiate a statistical investigation, as described in Figure 1.

Eighty students were enrolled in third grade, and every student answered the survey question. The data collected can be found in ThirdGradeSports.csv. Before proceeding to analyzing these data, we can identify features of the data by introducing key statistical terms.

The **observational units** are the entities on which data are recorded. They are the objects or individuals participating in a study. One can think of the observational units as what we measure. For this investigation, because we are taking a measurement of the students, the students are the observational units. The students' preferences are then recorded as data. In our investigation, we have 80 observations; each represents a student's response to the survey question.

A **population** is the entire collection of observational units that are of interest for a statistical investigation. For this investigation, the 80 students represent the population of third graders at the school. It is important to note that the investigative question defines the population of interest. In our case, the question "What sports do third grade students at our school prefer? refers to the fact that we are interested in the entire third grade class in our school—all 80 students. Because our survey was administered to the entire population of third graders, we conducted a **census**. A census gathers information from every member of the population of interest.

Our survey asked students to choose which sport they preferred to play at school. The **variable** we are interested in is the students' sport preferences. A **variable** is a characteristic or an attribute that describes the observational unit. A variable can fluctuate from one observational unit to the next. Variables can be either quantitative or categorical. A variable is **quantitative** if it represents a measurable quantity. The measurable quantity is measured numerically. For example, measurement of a person's height, student test scores, or a country's population size would all be considered quantitative variables. However, not all variables that have numbers as values are quantitative. Consider, for example, a person's zip code. Although a zip code is numeric, it is not a measurable quantity. A zip code simply identifies a person with a specific location of residence. This identification places a person in their respective zip code category. A variable is **categorical** when it assigns the observational units to a particular group. Therefore, a zip code is categorical. Other examples of categorical variables include a student's gender, a student's race, or the breed of a dog. In this investigation, the students' sport-preference variable is categorical. This is because each student has assigned themselves to a favorite sport (category). To prompt the analysis of these data, we can ask questions such as (a) how many students preferred each sport?, (b) what percentage of students preferred each sport?, and (c) what is the modal sport?

Categorical data can be summarized in a **frequency table**. A frequency table is a table that displays the categories of the variable and the number of observational units that identified with that category. It is important to understand that the actual data value is the category in which a student is classified. The frequencies are not the data. Instead they are summary numbers of the category data. The frequency table can help answer questions about the number of students who preferred each sport and the modal sport category.

What is your favorite sport to play at school?	Frequency
Basketball	25
Football	10
Handball	16
Soccer	14
Tag	15
Total	*80*

We see that basketball is the preferred sport for 25 of the 80 third graders, while football has 10 students, handball has 16, soccer has 14, and tag has 15 students. While handball, tag, and soccer are popular among the third grade students, basketball is the **modal** sport—that is, it is the sport most frequently selected among the third graders. Basketball is favored by approximately 31 percent of the students, while the second-most-favored sport, handball, has only 20 percent of the votes.

By looking at these analyses, we can then ask the interpretive question: What type of equipment should the administrators prioritize? We can see that basketball is the preferred sport of the third grade class. Thus, it would be reasonable that one conclusion the school administrators could make would be to prioritize the purchase of new basketballs and basketball hoops. If any money is leftover, then handball supplies should be purchased next.

The previous investigation demonstrates how a statistical investigation is guided by an investigative question. The investigative question "What sports do third grade students at our school prefer?" was answered through the **collection of data** from a class census, with a categorical variable identifying the favorite sport of each student. These data were collected using a survey question. The **analysis** included the summary of the data into a frequency table, followed by asking analysis questions such as "What is the modal category?" The **interpretation** of the frequency table enabled the administration to answer the investigative question directly. The investigation also highlights some important terminology, namely, *populations, census, types of data, observational units*, and *variables*.

INVESTIGATION SUMMARY:

The main concepts developed in the third grade sports investigation are:

1. The **statistical investigative process** includes formulating investigative questions, collecting or considering data, analyzing data, and interpreting data.
2. A **population** is the entire set of items, events, people, objects, etc. that are of interest for a posed investigative question.
3. An **observational unit** is the unit that the data describe. It is the object, person, group, item, etc. that we measure.
4. A **variable** is a characteristic or an attribute that describes the observational unit. The value of a variable can vary from one observational unit to the next. A variable can be quantitative or categorical. It is **quantitative** if it represents a measurable quantity that is measured numerically. It is **categorical** when it assigns the observational unit to a particular group.

Follow-Up Questions

1. Identify and describe the **populations** of interest for the following investigative questions of interest:
   a. What is the typical mathematics test score of students in the Mira Beach School District?
   b. What is the typical mathematics test score of third grade students in the United States?
   c. Do seventh grade students in California who tend to study late at night tend to have bad grades?
   d. Are 10-year-old kids in the United States more likely than 10-year-olds in Italy to have participated in organized sports?
   e. Can you predict how far a kangaroo in Australia can jump based on the kangaroo's height, age, and weight?
2. Identify the **observational units** in the following questions:
   a. Do students in the fifth grade class at Placid Elementary School prefer to read, explore mathematics, or neither during their free playtime each day?
   b. Does a school's impact on the environment differ between high-poverty areas and low-poverty areas in the Los Angeles Unified School District?
   c. Do teachers with more content knowledge in the Memphis City School District tend to give more open-ended tasks in the classroom?
   d. Within the different regions of the United States, what types of professional development are offered to teachers to support their pedagogy?

3. Determine what **variable** would be measured from the following survey questions, and determine whether the variable would be quantitative or categorical:
   a. What is your favorite professional men's basketball team?
   b. How many times have you traveled outside the country?
   c. In what month were you born?
   d. What is your height to the nearest inch?
   e. What is your telephone area code?

Case Studies: Seeing the Statistical Problem-Solving Process in News Articles

Every day, we are consumers of data. From commercials comparing one cellular-phone service with another, to National Public Radio segments discussing recently conducted studies, to news about polls gathering information on public opinion, we are presented with data in all facets of our lives. As users of data, it is important for us to be informed on how to interpret statistical ideas. We need to stay current and think critically about the world around us. To evaluate the conclusions presented in news outlets, we must be able to understand the studies that are referenced in these outlets. In particular, when we read about a study in an article, it is important to be able to recognize the investigative process that was carried out by the researchers in order to think critically about the results. We next present two news articles as case studies and attempt to detect the investigative process being discussed in articles reporting on studies.

Case Study 1: Mathematics and Exercise

Goals of this case study: Provide examples of statistics encountered in everyday life, connect the aspects of a statistical study to the statistical investigative process, and explain the difference between observational and experimental studies.

The article "Math-letes rule! Fit, Healthy Kids Do Better in School, Especially Math" was published in August 2015 on the CNN website. The article can be found here: www.cnn.com/2015/08/31/health/fit-kids-better-math/index.html.

The article describes a study that investigated the connections between kids' fitness levels and their performance in mathematics and English.

Read the published article and outline the statistical investigative process by:

a. identifying the investigative question,
b. describing how the data were collected,
c. describing how the data were analyzed, and
d. characterizing what interpretations were made from the data.

The article "Math-letes rule! Fit, Healthy Kids Do Better in School, Especially Math" discusses an **observational study** in which 9- and 10-year-olds were tested in mathematics, reading, and physical activity. This study is called an observational study because it is a study where researchers observe the observational units and measure them in some way (on some variables) without attempting to influence the outcome (as compared with an experimental study). The overall scope of the research was to examine the connection between the mind and the body. The investigative question the researchers set out to answer could be articulated as follows: Is there an association between mathematics test scores of 9- and 10-year-old children and the physical activity of a child? The physical activity was measured as the amount of time the child could run on a treadmill.

The investigative question indicates an interest in understanding whether there is an association between test scores and physical activity. The news article alludes to three different variables collected for each student. Because the students are the ones being measured, they serve as the observational unit of analysis in the study. The variables measured are: (1) the mathematics test score, (2) the reading test score, (3) the amount of time students could run on a treadmill while the treadmill was being increased by 3% grade increments every two minutes, and (4) the amount of gray matter in the students' brains, as shown in an MRI. Students stopped running when they were considered exhausted according to their oxygen uptake and respiratory exchange ratio, measured every 20 seconds. All four of the variables discussed are quantitative variables.

When thinking about these data, it may be helpful to picture a spreadsheet in which each row represents a student. To organize the data, there could be four columns in the spreadsheet representing the mathematics test score, the reading test score, and the time the child was able to run. The spreadsheet could look like the following table:

Observational Unit	Math Test Score (out of 100)	Reading Test Score (out of 100)	Run Time (min)	Amount of Gray Matter in the Brain
Student A	71	50	9	volume value
Student B	92	30	7	volume value
Student C	56	70	12	volume value

Forty-eight students were included in the study. Further internet investigation led us to the online journal article (https://journals.plos.org/plosone/article?id=10.1371/journal.pone.0134115), which describes more details of the study, such as where the study took place, how students were recruited to participate, and the researchers' screening process in selecting the 48 students. The students participating in the study were a **sample** of volunteers from the larger population of interest in the study, which was all 9- and 10-year-olds. A **sample** is a subgroup of a population of interest or a selected group from a population. A **population** is the set of all people or objects of interest in a study.

From the news article, we can distinguish each phase of the statistical investigative process in the following way:

(a) The study investigates the question, what is the association between mathematics test scores and the physical activity of a 9- or 10-year-old child?

(b) The data were collected through a sample of 48 9- and 10-year-old children who were tested on endurance on a treadmill, a mathematics test, and a reading test. In addition, the kids were given an MRI to measure the tissue in their brains.

(c) The data were analyzed by examining links between fitness and mathematics achievement. The specific statistical analyses conducted were documented in the full study, accessed by clicking through the links provided in the news article. The analyses included descriptive analyses, independent t-tests, and a multivariate analysis of variance.

(d) The analyses led to the interpretation that higher levels of fitness were associated with higher mathematics achievement. In addition, higher levels of fitness were associated with less gray matter in the front of the brains on the MRI scans. This gray matter is the area of the brain that is attributed to working memory, cognitive flexibility, and the ability to tune out distractions, all which are important skills in mathematics.

CASE STUDY SUMMARY:

The main concept developed in the mathematics and exercise study are:

It is helpful to dissect the investigative process of a case study in order to be informed how data are being reported. Through a careful reading of a news article and reference to the original study on which the news outlet reports, the statistical investigative process can be identified.

Case Study 2: Grocery Shopping and Healthy Eating

Goals of this case study: Provide examples of statistics encountered in everyday life, connect the aspects of a statistical study to the statistical investigative process, and explain the difference between observational and experimental studies.

The news podcast "How Partitioned Grocery Carts Can Help Shoppers Buy Healthier Foods" was published in May 2015 and discusses the study examined in the article "Partitioned Shopping Carts: Assortment Allocation Cues That Increase Fruit and Vegetable Purchases." The article can be found here: https://www.npr.org/2015/05/26/409671975/how-partitioned-grocery-carts-can-help-shoppers-buy-healthier-foods; a related article is here: www.npr.org/2015/09/10/439104239/can-grocery-carts-steer-consumers-to-healthier-purchases.

Read the published transcript from the podcast and outline the statistical investigative process by:

a. identifying the investigative question,
b. describing how the data were collected,
c. describing how the data were analyzed, and
d. characterizing what interpretations were made from the data.

The podcast transcript indicates that customers whose carts had large sections labeled for fruits and vegetables tended to purchase those healthier foods. As you reflect on the article, ask yourself whether you trust the findings. Why or why not?

The observational units were the individual people shopping at the store. In looking at the original published study, we find that 171 shoppers participated in the study, 75 shoppers used the partitioned carts, and 96 shoppers used normal carts. This was an experiment, and shoppers were randomly assigned the different types of carts. The two variables measured were the type of cart shoppers received (categorical) and the amount of fruits and vegetables purchased (quantitative). The amount of fruits and vegetables purchased was measured by the study as the dollar amount spent on fruits and vegetables. The population of interest was all individuals who shop for groceries. We can recognize the statistical investigative process in the following way:

(a) The study poses the following investigative question: Do partitioned carts encourage people to spend more money on fruits and vegetables compared with nonpartitioned carts?

(b) The data were collected in a grocery store. A total of 171 people participated in the study and were randomly given a type of shopping cart. The observational unit was each individual shopping for groceries. The study noted the type of cart they used and the dollar amount they spent on fruits and vegetables. These were categorical and quantitative variables, respectively.

(c) The data were analyzed by examining whether people with the partitioned carts spent more on fruits and vegetables than those with regular carts. To examine the impact of the type of cart, a regression analysis was performed.

(d) The study interpreted the results by noting that people with the partitioned carts were more likely than those with normal carts to spend more money on fruits and vegetables. The interpretation was that the type of cart used does in fact affect a person's shopping patterns.

CASE STUDY SUMMARY:

The main concept developed in the grocery shopping and healthy exercise study are:

It is helpful to dissect the investigative process of a case study in order to be an informed consumer of data. Through a careful reading of a news article and reference to the original study on which the news outlet reports, the statistical investigative process can be identified.

These case studies provide examples of how we are confronted with statistics in our daily lives. As consumers of data, it is important for us to know how to interpret informational articles and statistics in the news. By identifying the statistical investigative process, we are able to connect why we study statistics to real world examples, making our content knowledge applicable and meaningful.

As described in the introduction of this book, a primary purpose of the book is to provide a guide to prepare teachers as envisioned in the *Statistical Education of Teachers (SET)* report. The material covered in this unit directly aligns with specific *SET* guidelines, as well as current state standards. Alignment matrices are provided that illustrate which standards and guidelines each of the investigations and case studies in the unit cover. In general, *SET* adapts the *GAISE* model for the statistical thinking process. This unit introduces this process, which will be carried out throughout the book. Overall, statistics content should be introduced through investigations that follow the statistical process.

Follow-Up Questions

1. Read the four news articles linked in the table and outline the statistical investigative process by:
    a. identifying the investigative question,
    b. describing how the data were collected,
    c. describing how the data were analyzed,
    d. characterizing what interpretations were made from the data.
2. For each article, reflect on what additional information might be helpful for the article to include for the reader to get a better understanding of the study.
3. Is there anything you would do differently if you were conducting the study?

Case Study Name	Source
Migraines	https://www.fda.gov/news-events/press-announcements/fda-approves-new-treatment-adults-migraine
Busyness	www.huffingtonpost.com/entry/being-busy-is-actually-better-for-your-brain-study-finds_us_573c7f69e4b0ef86171ccab2
Lottery	http://news.health.com/2016/01/27/people-gamble-more-when-they-think-things-are-going-their-way/
Voice Changes	www.npr.org/sections/health-shots/2015/01/05/371964053/how-a-position-of-power-can-change-your-voice

Find a news article of interest to you and answer questions 1–3. Studies can be found via a Google search. Additionally, many can be found at the following website: www.npr.org/people/137765146/shankar-vedantam.

References

Arnold, P., and C. Franklin. (2021). What makes a good statistical question? *Journal of Statistics and Data Science Education, Vol. 29, 1.*

UNIT 1B:
The Role of Questioning in Statistics

As discussed in the previous unit, statistics is an investigative process guided by four components: Formulate Statistical Investigative Questions, Collect/Consider the Data, Analyze the Data, and Interpret the Results. While the process explicitly mentions questions in the Formulate Statistical Investigative Questions component, in fact, questioning plays an important role in all of the components. Namely there are questions that motivate a study, questions that motivate the need to collect data (investigative questions), questions that produce data, questions that prompt analyses of the data, questions that are focused on the interpretation of results, and interrogative questions that are asked as checks and balances throughout the whole process. In essence, questioning can be used throughout the components to guide the investigation and offer insights into each step of the process. The purpose of this unit is to illustrate different types of questions and show how they can help create and guide rich investigations.

Investigation 1B.1: Developing Investigative Questions

Goal of this investigation: Illustrate how questioning can guide an investigation through each of the four components—formulate questions, collect data, analyze data, and interpret results.

As part of Ms. Johnson's role as an administrator at a large middle school of 880 students, she is tasked with the job of deciding where to allocate school money and effort. At the beginning of the school year, the superintendent of her district provides her with a list of tasks that she will be responsible for throughout the year. She is expected to ensure that she makes a data-driven decision for every task.

Ms. Johnson's Task List:

1. Decide on the type of healthy and desirable food offerings for the cafeteria.
2. Help teachers determine whether they are meeting their teaching goals set in their professional development.

3. Decide what color to paint the school next year.
4. Determine what teachers need in order to feel they are being given adequate support to deliver the current mathematics standards.
5. Identify which subgroups of students might be struggling or succeeding within a class.
6. Determine whether having recess before math helps or hinders student focus.

For each task, (1) construct an initial investigative question and (2) create a data collection plan. Once that is complete, briefly describe the type of analyses one would conduct that would allow one to interpret the results and answer the investigative question.

Constructing an initial investigative question can be quite challenging, but writing statistical questions is a necessary skill for teachers to acquire in the interest of mastering the investigative process. For teachers to effectively guide a statistical investigation in a classroom, they need to be able to write and pose investigative questions. Writing a statistical investigative question first entails understanding the scenario to be researched and understanding the overall question motivating the study. For example, task 1: What types of food can the cafeteria provide that are both healthy and desirable choices? Answering such a question requires a decision about food offerings; therefore, it would be important to discover the food preferences of the students who would be purchasing the food. Understanding the underlying motivating question means focusing on the point of interest. Once the point of interest is identified, one can phrase a more specific investigative question geared at uncovering the issue. The question must be posed in a manner that requires data to be collected. The investigative question posed must anticipate variability in the data. For example, because the food preferences of middle-school students will be different depending on the student, possible investigative questions could be "What are the healthy food preferences of students?" or "What healthy foods do students typically prefer to eat?" Both questions require preference data from students to be collected. Although both questions ask about students' overall preferences , answering the second question requires considering how preferences will vary from student to student and trying to understand what a "typical" preference would be. Additionally, the task focuses on healthy food offerings in the school cafeteria. Thus, the following phrasing of an investigative question would be favored:

What healthy foods do our middle-school students prefer to eat at school?

Arnold (2013) developed criteria that can be used to critique statistical investigative questions that are posed. Using the investigative question, we can work through the criteria to see if the question is a good one.

1. **Is the variable of interest clear and clearly defined?** Yes, the variable of interest is healthy foods students prefer to eat.
2. **Is the group or population that we are investigating clear?** Yes, the question is about the students in the middle school where Ms. Johnson works: "our middle-school students."
3. **Is the intent of the question clear?** Yes, the investigative question is a summary-type question; we will summarize and describe the data collected to answer the investigative question.
4. **Can we answer the question with the data we can collect?** Yes, we can ask students about food preferences.
5. **Is the question about the whole group?** Yes, our question considers all of the responses from all students; it is not just asking about one category, for example.
6. **Is the question interesting and/or purposeful?** Yes, Ms. Johnson has to answer the question as part of her tasks for the year.

Ms. Johnson is interested in getting an idea of students' opinions about what they prefer to eat, so she decides to construct a survey. It should be noted that survey questions are not investigative questions. Survey questions are developed and posed for the purpose of data collection, while investigative questions set the stage for the entire process. Once an investigative question is posed, then appropriate data collection plans must be developed. A survey question is a question that produces data. In this sense, a survey question is a type of question that lives in the Collect/Consider the Data component of the investigative process. There are many ways to construct an appropriate survey question to help answer this investigative question. Here are two possible options.

Survey Option 1: To focus the survey, Ms. Johnson decides to provide a list of all healthy food options and ask students to rate each food item with their likability score for eating the item at school.

Rate each item on a scale of 1–3, where a rating of 1 means you do ***not*** *like to eat this item at school, 2 means you* ***sometimes*** *like to eat this item at school, and 3 means you* ***really like*** *to eat this item at school.*

Food Item	Circle a Rating		
Apple	1	2	3
Smoothie	1	2	3
Milk	1	2	3
Salad	1	2	3
Carrots	1	2	3
Yogurt	1	2	3
Sandwiches	1	2	3

The variables then consist of the students' scores for each of the food items. The variables are ordinal (have order). It might seem appropriate to state that the scores are categorical because they represent *dislike, like,* or *really like,* but the numbers represent an ordered scale of likability. In this sense, the numeric value of the food rating represents the degree of likability, which is ordinal; therefore, we are going to consider it quantitative. To treat it as quantitative, we are assuming that the difference between the numbers on the scale is consistent, which seems plausible in this case. Therefore, the data spreadsheet for a few example students will look like the following:

Student's Name	Apple Rating	Smoothie Rating	Milk Rating	Salad Rating	Carrots Rating	Yogurt Rating	Sandwiches Rating
John	2	3	3	3	2	1	3
Sara	1	3	3	3	2	1	2
Shayne	1	3	2	2	2	1	3
Kyle	3	3	3	2	2	2	3

Once the data are collected, the analysis seeks to provide understanding of the variability present in the data, while also trying to answer the investigative question. The following questions can be posed to help guide the analyses:

1. How do the food item scores compare with one another?
2. What is the mean score for each of the food items?
3. What is the food item that has the least/most variability in scoring?

These analysis questions offer guidance in the data analysis stage for the statistical investigative process. By quantifying and describing features of the variables, Ms. Johnson can then answer the initially posed statistical investigative question.

Survey Option 2: To focus the survey, Ms. Johnson decides to have students select their favorite food to eat at school from a list of foods.

Select the food item that you most prefer to eat in the cafeteria:

Food Item
Apple
Smoothie
Milk
Salad
Carrots
Yogurt
Sandwiches

In this survey, the variable is the students' favorite food, which is a categorical variable. Therefore, the data spreadsheet for a few example students might look like:

Student's Name	Favorite Food
John	Smoothie
Sara	Apple
Shayne	Milk
Kyle	Smoothie

Using these data, Ms. Johnson can consider how the food items compare with one another by answering the following guiding analysis questions:

1. What foods do middle-school students prefer to eat at school?
2. What is the most/least preferred food to eat at school?
3. Were the foods generally favorable or not favorable?

These three guiding questions for the data analysis stage will help answer the investigative question by describing features of the variables.

We now shift our interest to the next task on Ms. Johnson's list, task 2: Help teachers determine whether they are meeting their teaching goals set in their professional development.

The school district has been implementing a professional development program to help teachers improve their mathematics teaching. As part of the program, each teacher has to set a goal for their teaching. The goal must be based on some type of change in pedagogy that the teacher is trying to implement in their class. For example, one of the teachers, named Maria, decides that getting students to ask questions as they are solving mathematics problems is a practice that should be fostered. Maria believes that questioning can help students foster mathematical habits of mind (e.g., perseverance, critiquing, reasoning, modeling), which may lead to higher achievement. In her teaching, Maria's strengths lie in modeling questioning, and in encouraging students to ask themselves questions when they solve problems. Now, her goal is to investigate whether students' questioning does in fact help increase students' achievement. Her motivating research question is: Is there an association between student questioning while solving mathematics problems and their achievement in mathematics?

She poses the following statistical investigative question to motivate her data collection:

How do test scores on a mathematics test compare between students who ask questions and students who do not ask questions?

Maria decides that she will study her entire Math 1 class. Her unit of observation will be the student. Maria is interested in examining the link between student achievement on a mathematics test, a quantitative variable, and whether or not they asked themselves questions while taking the test, a categorical variable. To help her collect the data, she poses two data collection questions:

- Does a student ask questions?
- What is a student's score on the math test?

She decides to collect the data by giving a test and asking students to note whether or not they are asking questions as they take the assessment. To measure whether or not students are asking questions, she creates an assessment that provides a section titled "Question Space" next to each problem. This section provides space for students to jot down their questions as they go through each mathematics problem. As Maria looks over the students' tests, she records whether or not the students utilized the question space in the way she has been modeling. The data spreadsheet will then be as follows:

Student's Name	Test Score	Questioning (yes/no)
John	89	no
Sara	78	yes
Shayne	82	yes
Kyle	70	no

Using these data, Maria can then examine if questioning helped improve test scores. Some guiding questions for the analysis phase could be:

1. What is the mean student score on this test for students who ask questions and for those who do not? What is the difference in the mean scores?
2. How much variability is present in test scores for students who ask questions and for those who do not? (i.e., do students who ask questions generally score similarly or differently?)
3. How do test scores of students who use questioning compare with those who do not use questioning? Can a graphical display be created that shows the test scores and how often each score was given for the students who use questions and those who do not? Do the graphs show that these two distributions are different? In what way?

After considering the guiding analysis questions, Maria could interpret if questioning led to an increase in test scores. Her interpretation could be guided by the following questions: Is the difference between the two distributions meaningful? Is the difference large enough to matter?

The analysis and the interpretation of results parts of this investigation will be carried out fully in subsequent units.

INVESTIGATION SUMMARY:

The main concepts developed in the developing investigative questions investigation are:

1. A statistical investigative question is a question that requires one to collect data. Investigative questions guide investigations.
2. Good statistical investigative questions specify answers to the following questions:
    a. Is the variable of interest clear and clearly defined?
    b. Is the group or population that we are investigating clear?
    c. Is the intent of the question clear?
    d. Can we answer the question with the data we can collect?
    e. Is the question about the whole group?
    f. Is the question interesting and/or purposeful?
3. Questioning is used throughout the statistical problem-solving process in order to guide data collection, analysis, and interpretation, and to interrogate all stages of the investigative process.

Follow-Up Questions

The possible statistical investigative questions and data collection plans for tasks 1 and 2 have been completed. Now, consider task 3: Decide what color to paint the school next year.

1. Two school administrators propose the following different investigative questions and approaches to answering their investigative questions. Which approach will provide the best guidance to Ms. Johnson in determining the paint color of the school? Answers should be argued based on the data collected and the possible implications of their analyses. Discuss the pros and cons in each approach. If neither approach is satisfactory, design your own approach and discuss why it is better.

a. **Mr. Washington's approach**: Mr. Washington believes that to feel ownership of and connection to a school, students must have a say in how the school looks. For the school's upcoming paint job, Mr. Washington believes that student input should be solicited to help make the decision. He poses the following investigative question:

What is the favorite building color of people in the school?

To answer this question, Mr. Washington proposes to survey all students at the school and ask them the following survey question:

What is your favorite building color?

The data will then be analyzed to see which color prevails among the students. Mr. Washington then proposes to choose the color based on the most commonly chosen answer by the students.

b. **Ms. Lopez's approach**: Like Mr. Washington, Ms. Lopez also wants to solicit opinions from the students at the school. Ms. Lopez poses the following investigative question:

What color do students typically prefer school buildings to be painted?

To answer this question, Ms. Lopez proposes to survey all the students at the school and ask them the following survey question: *Select the color you would most like to see the buildings at the school be repainted: gray, cream, terra-cotta, blue, or green.* The data will then be analyzed to see which color got the most votes among the students. Ms. Lopez then proposes to choose the color based on the most commonly chosen answers by the students.

Consider task 4: Determine what teachers need in order to feel they are being given adequate support to deliver the current mathematics standards.

2. Create two different initial investigative questions and create a data collection plan for each. Compare and contrast the approaches for best determining the course of action for the task.

Consider task 5: Understand which subgroups of students might be struggling or succeeding in a class. As a first step, clearly define the subgroups of students you are interested in comparing.

3. Create two different initial investigative questions and create a data collection plan for each. Compare and contrast the approaches for best determining the course of action for the task.
4. Construct your own initial investigative question to be used in an elementary-school classroom. (Note that the investigative question does not need to focus on the students. It could, for example, focus on ladybugs, sports teams, etc. What is important is that the question is appropriate for elementary-school students in the sense that the data collection plan and the analysis plan will be doable for elementary students). Create a data collection plan that could be used with a classroom of elementary students to answer your question. Then, state examples of analysis questions you could use to guide the analysis. How could the students use their analysis to answer the investigative question? Repeat this exercise for a middle-school classroom and a high-school classroom.

References for This Unit

Arnold, P. 2013. *Statistical investigative questions—An enquiry into posing and answering investigative questions from existing data* (Doctoral thesis). https://researchspace.auckland.ac.nz/handle/2292/21305.

Arnold, P., and C. Franklin. (2021). What makes a good statistical question? *Journal of Statistics and Data Science Education, Vol. 29, 1.*

UNIT 1C: Introduction to Distributions

In the previous units, we introduced the statistical problem-solving process and discussed the role of questioning in this process. In this unit, we will focus on analyzing data through visualization. Once our investigative question is posed and relevant data are collected, it is time to analyze the data. An important step in data analysis is creating appropriate graphical displays of the data that aid in visualizing and identifying patterns in the variability present in the data. For each variable in a data set, we consider the possible values the variable could equal and how often each of those values occurs.

The **distribution** of a variable describes the possible values the variable could assume and how often each of those values occurs. Oftentimes in advanced statistics, a distribution is also specified by an equation. The distribution of a variable can show which values of the variable are common and which are rare. This information can be helpful in understanding tendencies and extracting overall patterns from variables, such as typical values of the variable or how much the values of the variable vary.

The concept of a distribution and visualizing a distribution are difficult for students. Much research has shown that students tend to focus on only small aspects of graphical displays and fail to take in the larger picture of the entire distribution (e.g., Friel, Curcio, and Bright, 2001). For example, students will focus on a particular value, such as the most common value or the maximum value, and not consider the whole graphical display of the data. While a particular value may be important, overall patterns are equally, if not more, important. Thus, a focus of teaching distributions to students is getting them to note general patterns and possible trends.

The goal of this unit is to introduce distributions, graphical displays and visualization of distributions, shapes of distributions, measures of center, and measures of variability, as well as how to use these characteristics of distributions to draw meaningful conclusions about variables, while finding patterns in data.

Investigation 1C.1: School Dance

Goals for this investigation: Engage in the statistical investigative process and visualize data through distributions for categorical variables.

Suppose a middle school is interested in organizing a school dance. Some details about the event have yet to be decided and the committee of teachers organizing the dance seeks to gather student input in order to make some of the decisions. In particular, a decision needs to be made about the music offerings at the dance. Several different investigative questions could be posed to prompt data collection and help make a decision about the music to be played at the dance. For example, the following investigative questions would work:

1. What types of music are preferred by students at the middle school?
2. What music genres do students at the middle school prefer?

Each of these questions requires a statistical investigation to answer them: Data must be collected, analyzed, and interpreted in order to provide answers to each of these questions. Once a decision is made regarding the research question, a data collection plan must be formed. It is important to note that the plan for data collection must align with the goals of the study. As discussed in previous units, the population being studied is dictated by the investigative question posed.

For instance, if the scope of a study is about a particular group of students, then the population consists of all students in that group. If one is looking to understand what, for example, the sixth grade class thinks about a specific issue, then surveying the entire sixth grade class (taking a census) is a practical plan. However, if the population is all students at a large middle school and it would not be feasible to obtain information from each student, then one might need to collect data from a sample of students.

Suppose the following investigative question is chosen:

What types of music are preferred by students at the middle school?

For this investigative question, we could construct a survey that asks students their music preferences. The survey question to collect data could be:

What is your favorite type of music to have played at dances?

Responses to this survey question will produce data. The observational unit is the student and the variable defined is the type of music that the student identified as their favorite. This variable is categorical. Possible values of this variable are rap, rock, etc. Notice that the survey question is aimed at gathering information about each surveyed student's favorite type of music, while the investigative question is about the preferences of all students at the school. An issue with this type of survey question is that it may produce so many different responses from students that it could be difficult to analyze the data and make a decision about the type of music to be played at the dance. Because of this, it might be more useful to ask a closed question about music preferences. Such a survey question might read:

> Do you prefer that rap, rock, classical, hip-hop, or country be played at the dance?

The advantage of the closed question is that students are forced to choose among the five options. If the researcher leaves the question open-ended, then the researcher risks having too many different types of responses that would need to be further sorted into categories. A disadvantage of the closed question is that it could be limiting for the survey taker and possibly not align perfectly with students' preferences. A data collection plan also includes how the survey will be distributed (e.g., paper copies or electronic and at what time of day/school period), and how the overall data set will be organized (e.g., spreadsheet, Google document).

While it would be ideal to garner every student's opinion, the possibility of conducting a census of the school might not be practical. If this were the case, then one must adjust one's data collection plan to obtain data that would allow for the analysis and interpretation to occur. If you can't take a census, another option would be to take a sample from the population of students of interest. There are many ways to take samples from populations. For the purpose of introducing distributions, let's assume that the school administrators sampled 100 students; surveys were handed out to them in paper form, and the students were asked to complete the survey on the spot to ensure that every student selected responses to the questions. The sample responses given by the 100 students at a middle school are provided in SchoolDance.csv[2].

As mentioned previously, our first goal in analyzing the data is to visualize the distribution of the variable. We do this by creating graphical displays for the variable. Because students' musical preference is a categorical variable, we have several options for graphical displays to visualize the distribution of preferences. The distributions of categorical

2 These data are fabricated to be plausible and realistic.

variables can be visualized using a **bar graph**, a **pie chart,** or a **table.** We will look at each of these options to see how the different displays provide different types of information. We will start with a **table.**

Frequency Table Results for Music Preferences

Music Preferences	Frequency	Percent
Classical	3	3%
Country	22	22%
Hip-Hop	34	34%
Rap	11	11%
Rock	30	30%

The table shows the counts as well as the relative frequency of each category. For example, the count or the frequency of students who like rap is 11. The rap has a relative frequency of 11% of the total number of students. We can see that hip-hop is the most popular genre, garnering 34% of the student preferences; rock comes in second, with 30% of the vote; and country comes in third, with 22% of the vote.

Because we are looking to see if any particular music types are predominantly preferred by the students, a **bar graph** may be a good choice for a display. A bar graph shows the categories of the music preference variable on the horizontal axis, and the frequency with which the categories are chosen by the students on the vertical axis. Alternatively, a bar graph could display the music preference variable on the vertical axis, and the relative frequency on the horizontal axis as a side bar graph. A bar graph allows us to view differences in frequencies by merely comparing the heights of the bars. The following bar graph (frequency counts on the vertical axis) indicates hip-hop was selected as the music of choice for the school dance by 34 students, rock was selected by 30 students, country was selected by 22 students, rap by 11 students, and classic rock by 3 students. The **mode** of the data is therefore the hip-hop category. The mode of a categorical variable is the category that occurs the most often. When referring to the typical value in a categorical data set, we are alluding to the mode. In this case, we call hip-hop the **modal category,** indicating that this was the most popular music choice among students.

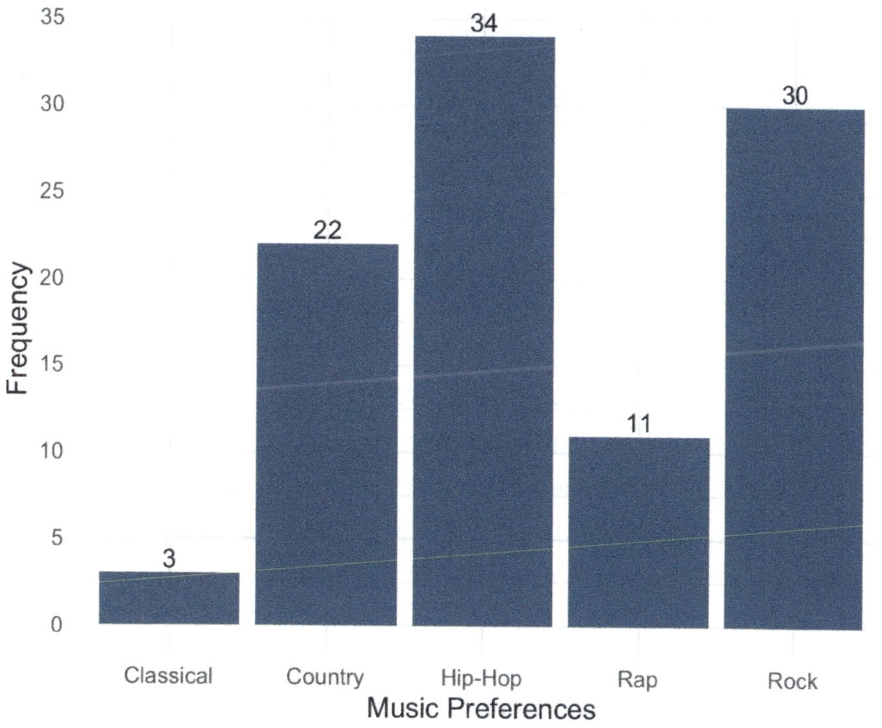

The third way to visualize the distribution of a categorical variable is through a **pie chart.** A pie chart displays each category as parts of the whole. The preference data can be visualized by this pie chart.

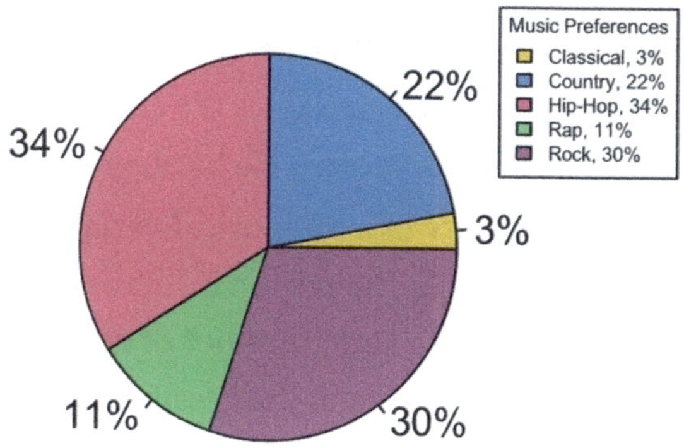

The pie graph shows that three main musical categories, hip-hop, rock, and country, dominate the preferences, while rap and classic rock were selected by a much smaller percentage of the sampled students. We can see that there is a fair amount of **variability** in these data, as no one category really dominates the rest. Variability, for categorical variables, describes how diverse the responses are across the response categories. As an exercise, one should envision what a pie graph might look like for a categorical variable as the differences among frequencies decrease and increase. This exercise could also be done with a bar graph possibly showing a clear preference of one category, say hip-hop, with more than 50%, and the other categories with small percentages. While there is no standard formal measure of variability for categorical variables noted in current state standards for school-level education, measures of variability for categorical variables do exist (see, for example, Kader and Perry, 2007).

Using the information from all three of the displays, we suggest that the music at the school dance consist of mostly hip-hop and rock, with some country songs incorporated throughout the dance. This investigation illustrated that we can display the distributions of categorical variables in three ways. Each type of display may offer different insights about the data. For example, when the variable has many categories, a pie chart might be difficult to read; thus, a bar graph or a table may be better options to view. When we want to visualize the categories as parts of a whole and compare the relative size of each category, a pie chart offers a very good visual. If the data have a very dominant category, the pie chart can easily draw attention to this fact. However, the bar graph would be the best option to see the modal category if each category had similar frequencies. A table is a worthwhile option when one wants to note the exact counts or relative frequency in each category. When visualizing the distribution of a categorical variable, it is important to consider which type of display is best for the variable at hand. It is preferable to look at multiple displays of distributions in order to explore all aspects of the variable.

INVESTIGATION SUMMARY:

The main concepts developed in the school dance investigation are:

1. The distribution of a categorical variable can be represented and visualized through a bar graph, a pie chart, or a table.
2. Many categorical variables are summarized by their modal category and described by how often each category occurs.
3. The variability of a categorical variable can be gauged by how much disagreement there is in the responses across the various categories. When there is a lot of agreement (a large number of responses in one category), then there is a little variability in the data. If all responses are in the same category, then there is no variability in the data. The more disagreement, the more variability.

Investigation 1C.2: Practice Test Scores

Goals for this investigation: Engage in the statistical investigative process, visualize data through distributions for quantitative variables, and measure possible variation in the data.

Mr. Garcia is a middle-school teacher. He is interested in understanding how his students are performing on a practice test for the state's standardized test administered at the end of the school year. He poses the following investigative question:

What are typical test scores on the practice test taken by Mr. Garcia's students?

Mr. Garcia has a total of 96 students in three classes. All of his classes have students with similar academic abilities, but he is unsure how they will perform on a standardized test. The students' results on Mr. Garcia's practice tests are given in PracticeTestScores.csv.

Mr. Garcia is interested in knowing the typical test score of students in his classes. The test score variable is quantitative; therefore, the graphical displays used in the previous investigation with a categorical variable are not appropriate. The distributions of quantitative variables can be visualized using **dotplots, histograms**, or **boxplots.** In this unit, we will focus on introducing dotplots (also sometimes referred to as line plots, but throughout this book will be called dotplots) and histograms. Boxplots will be discussed in the next unit, when we begin comparing distributions. To begin this investigation, we will use a **dotplot** to visualize the data:

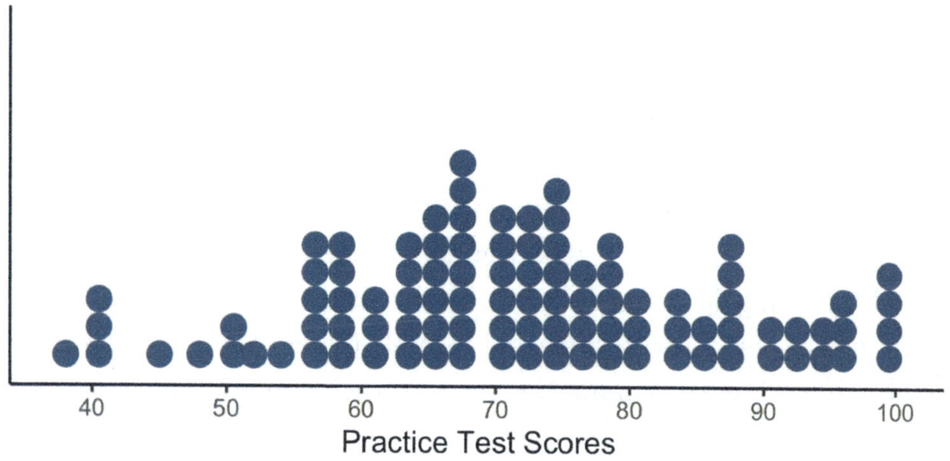

Each dot on the plot represents a student's test score. When describing the distribution of a quantitative variable, we first focus on describing its **shape, center,** and **variability.** This dotplot has a single-mounded shape centered on a test score of 70 points. The distribution appears to be approximately symmetric[3] about the test score of 70, with about the same number of scores on either side of 70. As you get further from either side of 70, the frequency of data points tends to decrease.

Because of the approximate symmetry of the distribution about the center and the single mound of the distribution, the **mean** (computed by summing all of the practice test scores and dividing by the total number of scores) is an appropriate measure of center for these

3 Note that if a distribution is symmetric, it does not necessarily mean that it is normal. A normal distribution is a very specific type of continuous distribution that has specific features and properties in addition to being symmetric.

data. If, for example, the distribution was not symmetric, then it might be more appropriate to use the **median** (computed by putting all of the test scores in ascending order and then finding the middle score of the data) to describe the center of the distribution.

Because we're using the mean to note the center of the data distribution, the **mean absolute deviation (MAD)** provides a good introductory measure of variation for the data along with the **range**. We can see that test scores vary from 39 to 100, thus giving a range of 61 points (**range = maximum value-minimum value** = 100-39). Thus, 61 is the maximum possible distance between any two scores. We can also conceptualize the variability as the average distance of the points from the center of 70 through the MAD. To describe the variability in this way, we consider the distance of each of the student's scores from the mean value of 71.71. For example, there are a few students who have scored 40 points on the practice test. These students are approximately 30 points away from the central value. On the other hand, there are also students who have scored 100 points. These are also approximately 30 points away from the central value. Overall we can consider the distances from the mean value for every student, then find the average (mean) of these distances. This is important because it will give us a numeric value to summarize the variability from the mean in the data set. This numeric summary is the **MAD**.

To be precise, we carry out the computation. To compute the MAD, we calculate the distance from each point to 71.71 (the mean). This can be quickly done with software, such as Excel, by setting up formulas in a spreadsheet. An image of such an Excel spreadsheet is below. The spreadsheet shows the formula for the distance from the mean as the absolute value of the score minus the mean for each observation:

	A	B	C	D	E	F
1	Practice Test Score	Distance from Mean	Absolute Distance from Mean		Mean of Practice Test Scores	71.71
2	40	-31.71	31.71		Total Distance from Mean	1108
3	40	-31.71	31.71		Average Distance from Mean (MAD)	11.541667
4	76	4.29	4.29			
5	96	24.29	24.29			
6	75	3.29	3.29			
7	75	3.29	3.29			
8	61	-10.71	10.71			
9	77	5.29	5.29			
10	50	-21.71	21.71			
11	78	6.29	6.29			
12	84	12.29	12.29			
13	72	0.29	0.29			
14	68	-3.71	3.71			
15	88	16.29	16.29			
16	59	-12.71	12.71			
17	54	-17.71	17.71			
18	70	-1.71	1.71			
19	56	-15.71	15.71			
20	67	-4.71	4.71			

The spreadsheet shows that in column B we compute the distance from 71.71 for each test score. This is because we are not concerned about whether the test score is below or above the mean; we are merely concerned about its distance from the mean test score.

Cell E1 shows the total distance, 1108, away from the mean for the entire group of 96 students. To find the average absolute distance from the mean in cell E2, we then divide the total by 96, the number of test scores. The average distance from the mean, calculated as 11.54 points, is the MAD. The MAD is equal to the SUM[(observation-mean)]/number of observations and given by the following equation:

$$MAD = \frac{\Sigma(X_i - \overline{X})}{n}$$

where X_i represents observation i, $\overline{X}$ represents the mean, and is equal to the number of observations.

The MAD provides a simple computable measure of the variation present in the data. It is the mean distance of the data points from the mean value of the variable. In the context of test scores, a MAD of 11.54 seems quite large. The MAD shows that, on average, the 96 test scores vary from the mean test score of 71.71 by 11.54 points. Consider the letter-grade differences for a student who received a test score of 11.54 points above the mean, compared with a student test score of 11.54 points below the mean. This grade differential would be approximately 23 points, thus largely shifting their grades.

Another measure of variability that is often used is the standard deviation. Teachers will use the MAD with their middle-school students, and the standard deviation will be introduced in the general high-school math curriculum or in AP statistics.

The **standard deviation** measures the typical distance of test scores away from the mean, but unlike the MAD, which uses the absolute value to make the differences positive, the standard deviation squares the differences (see Franklin et al. 2020 as a resource for teaching and learning of the MAD). After the squared differences are computed, they are averaged (dividing by n-1 instead of n), and then the square root is taken to find the standard deviation. Again, for these data, we can use an Excel spreadsheet and specific formulas to show the computation:

	A	B	C	D	E	F
1	Practice Test Score	Distance from Mean	Squared Distance from Mean		Total Squared Distance from Mean	20315.8336
2	40	-31.71	1005.5241		Total Squared Distance from Mean/(96-1)	213.85088
3	40	-31.71	1005.5241		Square Root	14.6236411
4	76	4.29	18.4041			
5	96	24.29	590.0041			
6	75	3.29	10.8241			
7	75	3.29	10.8241			
8	61	-10.71	114.7041			
9	77	5.29	27.9841			
10	50	-21.71	471.3241			
11	78	6.29	39.5641			
12	84	12.29	151.0441			
13	72	0.29	0.0841			
14	68	-3.71	13.7641			
15	88	16.29	265.3641			
16	59	-12.71	161.5441			
17	54	-17.71	313.6441			
18	70	-1.71	2.9241			
19	56	-15.71	246.8041			
20	67	-4.71	22.1841			

The standard deviation for these data is given by the following formula:

$$SD = \sqrt{\sum_{i=1}^{96} \frac{(X_i - \overline{X})^2}{96 - 1}}$$

The standard deviation is typically found by dividing by one less than the number of observations in the data set (n-1). The sample mean is used to find the distance of an observation from the mean that is the balance point of the distribution. Thus, these differences of the (observed - mean) only provide n-1 unique pieces of information as the last difference must be whatever allows the sum of the differences to be zero. It can be shown mathematically that if the standard deviation is divided by n, the sample standard deviation will underestimate the population standard deviation. For more on this topic, one can view the following video:
www.khanacademy.org/math/ap-statistics/summarizing-quantitative-data-ap/more-standard-deviation/v/review-and-intuition-why-we-divide-by-n-1-for-the-unbiased-sample-variance)

The entry in cell F3 shows the standard deviation as 14.63. While this is similar to the value of the MAD (11.54), it is slightly larger. This is because when squaring instead of taking the absolute value, the standard deviation is more affected by values that are further away from the mean. For example, consider the test score of 40 in cell A2. That is a low test score value in Mr. Garcia's classes, and it is very far from the mean. The squared distance from the mean for the score of 40 is 1005.53, which is very large. For this same test score value, instead of the squared distance, the MAD uses an absolute distance from the mean of 31.71, much smaller.

To compare how the MAD and the standard deviation treat far-away and close values to the mean, consider the value in cell A4, 76, a test score that is close to the mean. The MAD uses the absolute distance from the mean at 4.29, while the standard deviation uses the squared distance at 18.4. While the standard deviation distance value is still larger than the MAD value, it is not so much larger, unlike the case of the value of 40, seen in cell A2, which is far from the mean. The MAD considers a distance from a value to the mean in the absolute value, which is linear, while the standard deviation considers the distance between a value and the mean squared, which is quadratic; therefore, as an individual value gets further and further away from the mean, the standard deviation grows faster than the MAD.

Both measures of variability offer insight into the overall deviation of the data values from the mean. The standard deviation will always be greater than or equal to the MAD because it weights the extreme values more. In general, the MAD is introduced in middle school as a measure of variability for quantitative variables because its computation is more intuitive than the standard deviation. Students have an easier time understanding the idea of computing a direct average distance from the mean than understanding the formula for the standard deviation. The formula for the MAD is direct, while the standard deviation hides some underlying nuances that are difficult for students to grasp. However, because the standard deviation is computed by squaring the distances from the mean instead of taking the absolute value of the distances from the mean, it becomes easier to deal with in advanced settings when one needs to take derivatives and integrals. In these settings, dealing with an exponent instead of an absolute value is much preferred. For these reasons the standard deviation is taught as early as high school and is the standard measure of variability used in high school and beyond.

In summary, the answer to the investigative question is that the mean test score for Mr. Garcia's class is 71.71. The mean can be interpreted as the balance point of the distribution (see *GAISE II* for an example developing this idea). The shape of the distribution (roughly symmetric) makes the mean an appropriate choice for the typical value. The variability of the test scores is large, so although Mr. Garcia has many students that are well prepared for the test, he also has many students scoring poorly. The MAD and the standard deviation, valued at 11.54 and 14.55, show that on average, students were an entire letter grade off of the mean score. Ideally, to do well on the standardized test, Mr. Garcia's students need to increase their mean test score value and have a smaller amount of variability around the typical value. To have smaller variability, all students would need to score closer to the typical test score.

INVESTIGATION SUMMARY:

The main concepts developed in the practice test scores investigation are:

1. The distribution of a quantitative variable can be represented and visualized through a dotplot.
2. The distribution of a quantitative variable can be described by its shape, center, and variability.
3. The variability from the mean of a quantitative variable can be described by the MAD and/or the standard deviation. The MAD is a more intuitive measure of variability; thus, it is taught in middle school. The standard deviation is then taught in high school.

Investigation 1C.3: Companies in Town

Goals for this investigation: Engage in the statistical investigative process and visualize data through graphical displays for quantitative variables.

The mayor of Mira Beach, a small beach town, is interested in understanding the financial profits of companies in town and identifying potential factors that might lead to people becoming successful business owners in the future. More specifically, the mayor is interested in investigating two questions:

- What is the typical weekly profit made by companies in town?
- What is the typical educational level of company leaders in town?

The small town has a total of 98 companies. As part of their annual reports to the Chamber of Commerce, these companies have to report on demographic and background information for their CEOs and their profit earnings as average profit per week in dollars. The mayor of the town has access to these annual reports and desires to answer the aforementioned investigative questions. The data are included in AnnualReport.csv[4].

4 Data adapted from *Insurance Profits* data set, available on StatCrunch.

As part of the demographic data, the CEOs report their educational level. The levels are reported in the following manner:

1 = High school
2 = Professional degree (technology, nursing, etc.)
3 = BA/BS
4 = Master's (MBA, MS, MA, MFA, etc.)
5 = PhD or equivalent

As a first part of the investigation, the mayor of the town aims to investigate the answer to the following question:

- What are the typical weekly profits made by companies in town?

Because we are interested in understanding what the typical profits are, we would like to visualize the distribution of the data to see if there are specific values of weekly profits that occur more frequently than others or if the profits are centered on a specific value. The profits variable is a quantitative variable, so dotplots, histograms, and boxplots are appropriate graphics to visualize the data. We begin by making a dotplot of the profits.

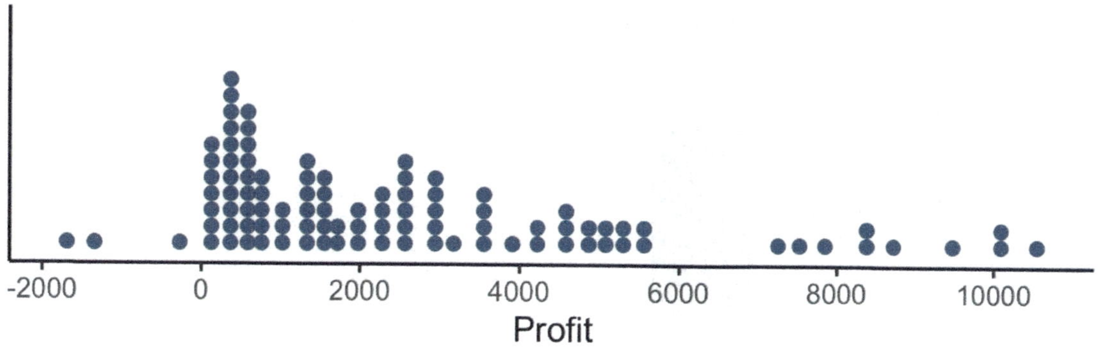

The dotplot provides a fine-grained visual of the average weekly profits in the town, because each dot represents the average annual weekly profits of one of the 98 companies. When looking at the distribution of a quantitative variable, we begin by describing the shape, center, and variability. A **skewed distribution** is one in which one of its tails is longer than the other. A tail refers to either end of the values on the horizontal axis of the distribution. We say this distribution is skewed right, because a large cluster of companies (85 out of 98) have weekly profits between $0 and approximately $5500, with a tail on the right end because a few companies posted weekly profits greater than $6000.

Based on the dotplot and attempting to draw a balance point line on the graphs, we would also estimate a typical profit for the companies in the town of about $2000. Because of the skewed right nature of the distribution, we would expect the median profit to be less than the mean profit, since the mean is pulled in the direction of the longer tail. This is because the mean uses each of the data values in its computation, so it will become larger if there are higher values present in the data. On the other hand, the median represents the halfway point in the data (the 50th percentile) and does not use each of the data values in its computation; therefore, it is not sensitive to extreme values of the variable. Because of this, when a distribution is skewed, the median is a more appropriate measure to describe a typical value.

The 98 companies had weekly profits that varied from as low as –$2,000 to as high as $10,500. Three companies had losses and 10 companies had profits in excess of approximately $7500. The range in the profits is approximately $12,000. Using software, we compute the standard deviation to be approximately $2683. This value can be interpreted as the typical deviation from the mean profit of the companies in the town. In the context of profits, this deviation seems moderate.

Because there are many different profit values, it might be useful to form profit "bins" (or intervals) of data that group similar profits. This type of graph is called a **histogram**; following is a histogram for the weekly profit data.

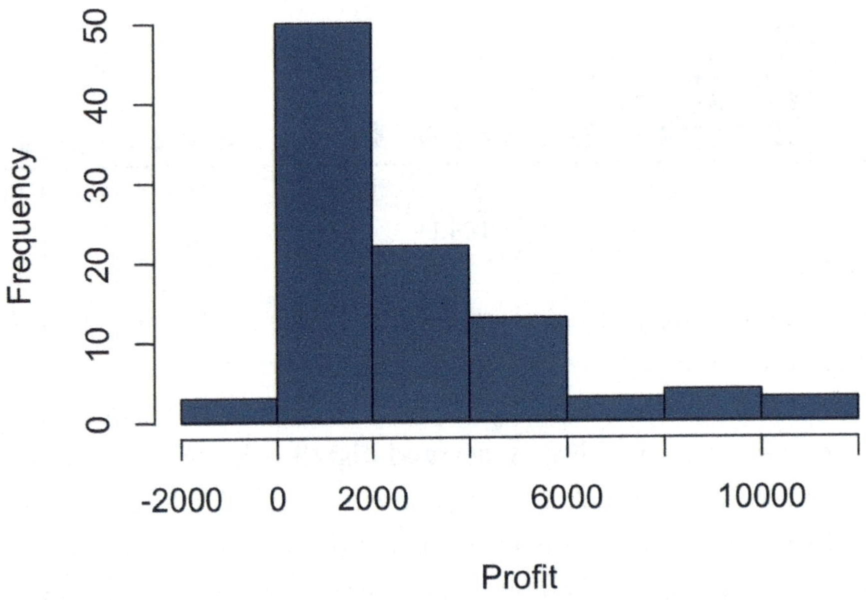

By looking at the histogram, we can attempt to identify a center. Because the distribution is skewed, the median is an appropriate measure of center. Without computing the median explicitly, the histogram suggests that the median lies somewhere around $2000. Overall, locating the center of a distribution when the distribution is skewed is difficult to do through visualization.

Students might want to describe the center as the interval of \$0–\$1000 because that is the interval with the highest bar and therefore the most data. There are two main issues with this reasoning. First, profits is a quantitative variable, and thus the preferred measures of center to use are the mean or the median (we usually reserve the modal category for categorical variables). Second, reporting the typical value to be within \$0–\$1000 does not take into consideration the shape of the distribution or the variability present. Typical values should be representative of the values of the variable in some way. Reporting the \$0 to \$1000 interval would not, for example, represent the profits of the companies that are larger. Therefore, a typical value that is somewhat more in the middle of the values would be better suited.

As noted previously, the center could be the mean or the median, depending on the shape and variability of the data. For these profit values, the distribution is skewed, so the median better represents the typical value of the data. It could also be the case that a distribution of a quantitative variable showed the mean or the median not near the most common values. For example, the following graph is considered a symmetric distribution with two clusters or humps (bimodal). The mean and the median would both lie in the middle of the two humps where no values occur; thus, they would certainly not be good descriptors of typical values. While the measures of center do not describe the most common values, they do describe the balance point and the halfway point in the data. In this sense, they could still be used to represent typicalness, although in this case, it would be best to describe the typical values as the two modal categories in each cluster, namely approximately 3 and approximately 19.

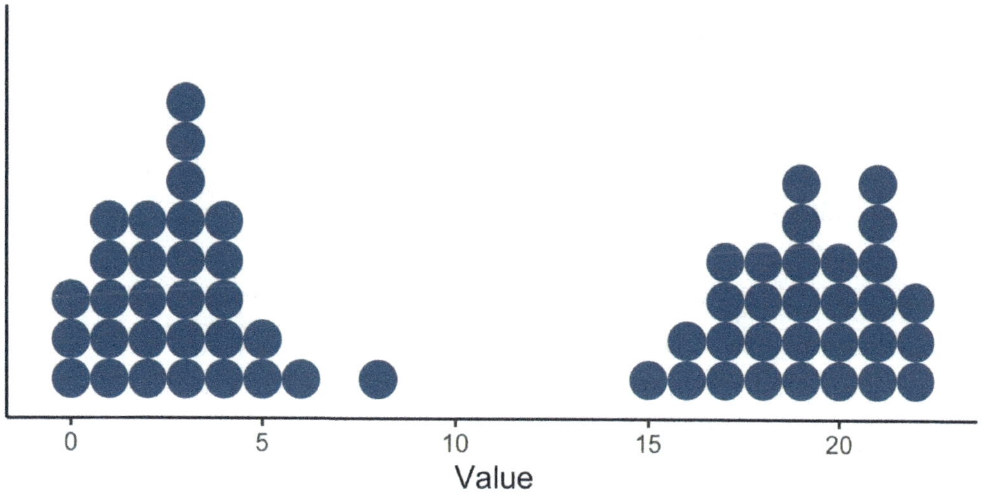

Dotplots can reveal more fine-grained patterns because they show the value of every company's profits, while histograms can uncover large patterns because the values are represented in larger bins. The histogram reveals that there are several companies that make large weekly profits around $8000 and above. The histogram makes clear that the shape of the distribution is skewed. Overall, the mean profit is driven up by these large-profit companies, which could be referred to as outliers (the concept of an outlier is discussed further in other investigations). Because the mean is sensitive to the skewedness in the shape of the distribution, we would guess the mean weekly profit to be around $3000 per week, even though we visually estimated the typical value to be around $2000. However, because the mean is sensitive to these few larger values, the median, which seems to be around $2000, is more representative of the center of the distribution. Examining the graphical displays and visually trying to note where the mean and the median lie, as well as noting the overall shape and variability present in the data, is a useful exercise. It is important to examine the graphical displays before jumping directly to computation to build intuition. After the visual interpretation, we can also compute the exact statistics:

Summary Statistics:

Variable	Mean	Median
Profit	2561.7	1644.5

The calculations show that our estimations based on the graphical displays were relatively accurate. The mean value of $2561 can be interpreted as the balance point or the "fair share" point of the distribution of profits. (See www.statisticsteacher.org/2020/11/12/the-mean-and-variability/ for a description of the mean as a balance point). We can imagine collecting all of the profits from the companies, and if we were to distribute these profits so every company in town had the same profit, then every company would have approximately $2561 in profits. The median also may be interpreted as a balance point as proposed by Lesser, Wagler, and Abormegah (2014). They show how to build a physical model to help conceptualize the mean and median as balance points of the data. See http://jse.amstat.org/v22n3/lesser.pdf.

At this point, we need to decide what value is the most representative of a typical profit value in this town. Based on the histogram visualization, we observed that the typical value would be around $2000. The median is $1645, which is close to our visual estimate. Because the higher profits of a few companies skewed the distribution to the right, the mean is larger than the median. The median is not as sensitive to these higher values and thus remains lower. In the case of a skewed distribution, the median provides a better representation of the typical value of weekly profits for the town.

INVESTIGATION SUMMARY:

The main concepts developed in the companies in town investigation are:

1. The distribution of a quantitative variable can be visualized through a dotplot and a histogram.
2. The distribution of a quantitative variable can be described by its shape, center, and variability.
3. A typical value for a quantitative variable can be summarized by the mean and median. The choice of which one represents the variable in a better way depends on the shape of the distribution.

In this unit, we introduced various graphical displays to visualize the distributions of categorical and quantitative variables. Each of these graphical displays is explicitly discussed in *GAISE, SET,* and current state standards. While they are all important, the sophistication of their features make some types of graphs more appropriate for different grade levels.

Categorical variables are a large focus in the elementary grades; thus, pie charts, bar graphs, and frequency tables are considered appropriate for the grade band. Quantitative variables and their graphs are more complex. Dotplots, histograms, and boxplots are introduced at the middle-school level. These suggestions align with the *GAISE* levels A, B, and C, which roughly coincide with elementary, middle, and high school. Similarly, the different ways to describe variability of quantitative variables are appropriate for different levels of students. The MAD is introduced in middle school, while the standard deviation is introduced in high school. For the measures of center, the mode is a categorical summary statistic and is recommended for elementary students, while the mean and the median are quantitative variable descriptors and are appropriate for upper elementary school and beyond. This unit provides an initial introduction to distributions and features of distributions for teachers of all grade bands.

Follow-Up Problems

1. Give examples of investigative questions that you could lead with your class related to the preference of fourth graders visiting a zoo, aquarium, or waterpark for a field trip.

The preference data from the fourth grade class at a school district was collected in the data set EndofYearParty.csv. Analyze and interpret these data to answer your posed investigative question using appropriate graphical displays.

2. The mayor in Investigation 1C.3, Companies in Town, articulated a second question to investigate: What is the typical educational level of company leaders in the town?
3. Give examples of statistical questions that you could pose to your class on foot measurement in fifth grade versus foot measurement in second grade.

   Foot measurement data from a fifth grade class and a second grade class was collected in the data set FootMeasures5&2.csv. Analyze and interpret these data to answer your posed investigative question using appropriate graphical displays. Give examples of statistical questions that you could pose to your class on foot measurement in fifth grade versus foot measurement in second grade.
4. Investigation 1C.1 walks through the computation of the MAD using Excel software. Compute the MAD for the foot measurements in both grade 5 and grade 2.

Reference for This Unit

Bargagliotti, A., Franklin, C., Arnold, P., Gould, R., Johnson, S., Perez, L., and D. Spangler. 2020. *Pre-K–12 guidelines for assessment and instruction in statistics education II (GAISE II)* report. American Statistical Association and National Council of Teachers of Mathematics.

Franklin, C., Kader, G., Jacobbe, T., and K. Maddox. 2020. The mean and variability from the mean. *Statistics Teacher.* www.statisticsteacher.org/2020/11/12/the-mean-and-variability.

Friel, S.N., Curcio, F.R., and G.W. Bright. 2001. Making sense of graphs: Critical factors influencing comprehension and instructional implications. *Journal for Research in Mathematics Education* 32: 124–58.

Kader, G.D., and M. Perry. 2007. Variability for categorical variables. *Journal of Statistics Education* 15(2).

UNIT 1D:
Comparing Distributions

The previous unit introduced distributions and explored the importance of using distributions to answer investigative questions. We saw that visualizing the distributions of variables helped us conduct analyses in a statistical investigation. Analyses can help us answer questions, such as what is a typical value or what music do students at a school prefer to play at a school dance. Distributions for quantitative variables can be summarized and described by their shape, center, and variability. Distributions for categorical variables can be summarized by their modal category, their frequencies, their relative frequencies, and the overall amount of disagreement and agreement between categories to quantify the variability of a categorical variable's distributions. This unit focuses on using distributions to draw comparisons between groups.

Investigation 1D.1: Student Sleep Patterns

Goals of this investigation: Engage in the statistical investigative process and compare distributions of quantitative data.

For humans to function properly, they need sleep. The National Institutes of Health (NIH) states: "Sleep plays a vital role in good health and well-being throughout your life. Getting enough quality sleep at the right times can help protect your mental health, physical health, quality of life, and safety." The topic of sleep and how it affects performance is discussed extensively in medical studies, cognitive studies, education studies, and sports performance studies. The table provides NIH recommendations for sleep amounts for people of different ages.

Age	Recommended Amount of Sleep
Newborns	16–18 hours a day
Preschool-aged children	11–12 hours a day
School-aged children	At least 10 hours a day
Teens	9–10 hours a day
Adults (including the elderly)	7–8 hours a day

www.nhlbi.nih.gov/health/health-topics/topics/sdd/howmuch

As can be seen in the table, middle-school and high-school students need approximately 10 hours of sleep per day. To date, there is much discussion around the large number of hours students spend doing homework and extracurricular activities. There is concern about how these activities are negatively affecting adolescent sleep habits and having potentially negative effects on their health. When students have homework and extracurricular activities for long hours, they are not able to rest their bodies, thus causing concern over the effects on healthy brain function, physical health, and the ability to function during the day. In this investigation, we will consider these ideas by examining students' sleep patterns.

In a local district, the principal of the high school is interested in comparing sleep patterns of students in the different high-school grade levels. The principal has observed homework pressures rise for students as they progress through high school, and he would like to understand if this progression affects students' sleeping patterns. He is particularly interested in sophomores, juniors, and seniors, because high academic pressure on students is typically reported in these years.

The principal selected 180 students each from the first periods of the sophomore, junior, and senior classes and asked them to record the amount they slept on a particular Wednesday night. The students were asked to report their hours in a shared Google document the next day during first period. Specifically, the exact prompt was: "At what time did you go to bed last night and what time did you wake up this morning?"

The principal then took their responses and found the difference rounded to the nearest tenth. The number of hours of sleep each student got the night prior was then recorded in StudentSleep.csv.

Using the data in StudentSleep.csv, answer the following investigative question:

How do the sleep patterns of the students in the different grades compare?

To answer this investigative question, let's first graphically display the distribution of the sleep patterns of the students in the three grades. Because the amount of sleep a student gets in a night is a quantitative variable, the three appropriate graphical displays to visualize the distribution of the data are the dotplot, the histogram, and the boxplot. The different types of displays will highlight the differences and similarities among each grade level.

Let's begin by comparing the dotplots of the students' sleep patterns among the three classes:

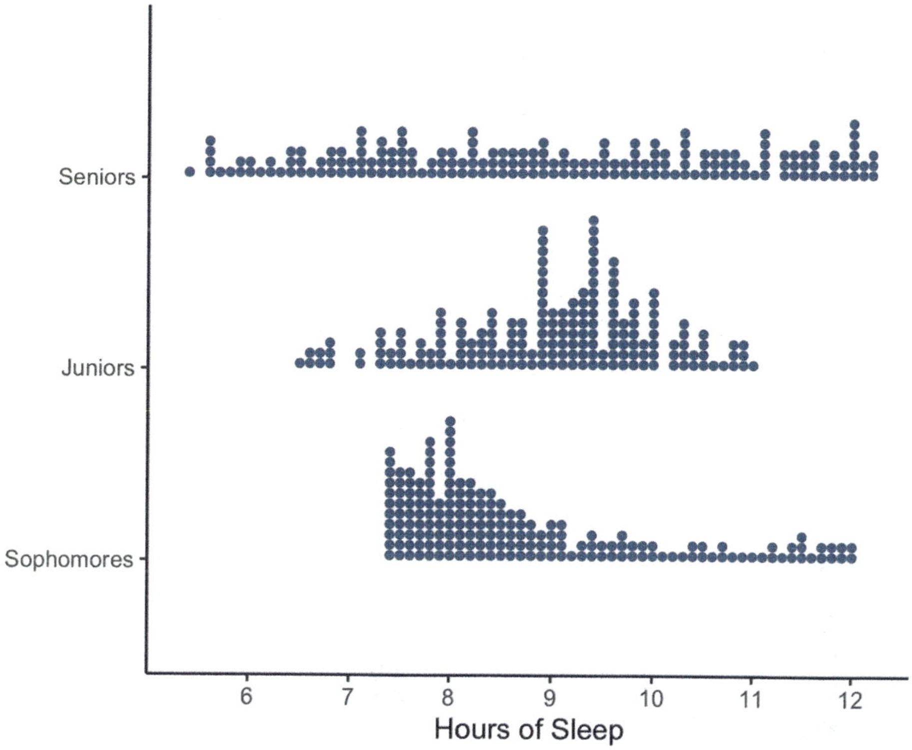

Approximately 36% of seniors, 17% of juniors, and 18% of sophomores got 10 or more hours of sleep. The **dotplot** for the seniors' sleep time appears to be evenly distributed between 5.5 hours to 12.2 hours. In other words, for every recorded hourly value between 5.5 and 12.2, the number of seniors sleeping for a given number of hours is approximately the same. Distributions with this property are said to be *uniform*. Note that this distribution is also reasonably symmetric with a central value located between 8 and 9 hours.

The juniors appear to have sleep patterns that are mound-shaped, with most of the juniors sleeping around 9.5 hours on the selected night. The further away the values get from 9.5 (in both directions), the less often they occur. This distribution is mound-shaped and somewhat symmetric. So, we say that the sample distribution of juniors' sleep patterns is approximately bell-shaped.

The sophomores' distribution illustrates that no sophomore slept less than 7.3 hours on the selected night. The majority of sophomores slept 7–9 hours, and very few students got more than 9 hours of sleep. The greater the number of hours of sleep, the fewer the number of students. We see that as the number of hours of sleep increases, the number of sophomores that sleep these higher hours trickles off. There are fewer students who sleep more hours compared with those who sleep less hours. Because the tail of the distribution falls to the right, we say the shape is right-tailed; therefore, the distribution is *skewed right*.

The **histograms** of the distributions show information similar to that in the dotplots. On a dotplot, the data corresponding to the individual are represented by a dot, but on a histogram, the data are grouped into bins. In the following histograms, the hours slept are grouped in bins that have a width of 0.5 hours (the widths can be adjusted as one prefers). The histograms show the skewness of the number of hours the sophomores slept, the mound-shaped distribution of the number of hours the juniors slept, and the uniformity of the number of hours the seniors slept. Because histograms group data, it is sometimes easier to see the overall shape of a distribution, but we lose the visualization of the individual observations as seen in the dotplot. For example, while the dotplot certainly illustrates the single-moundness of the juniors, this shape is easier to spot in the histogram.

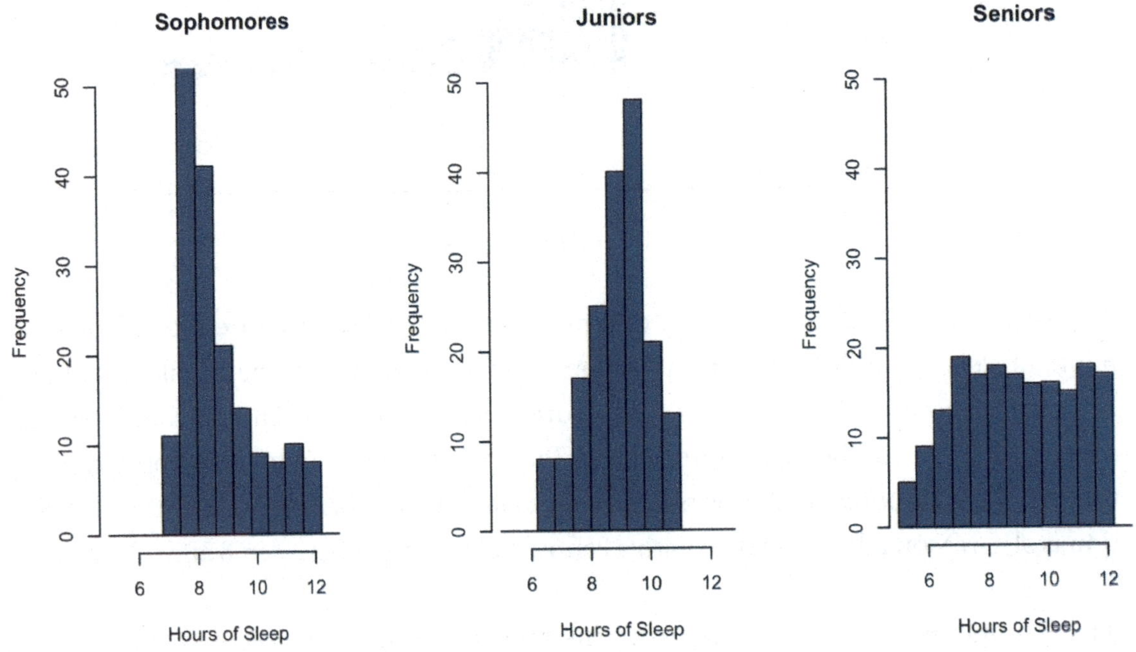

Another useful visualization for comparing the distributions is to use **boxplots** (also referred to as box-and-whiskers plots). Boxplots visualize measures of position of the data. They provide another way to look at the distribution by bringing to light patterns, such as comparisons of the different quartiles of the data and comparisons of the medians. Boxplots are visualizations of a distribution's five-number summary consisting of the minimum value, the first-quartile value (Q1 or 25th percentile), the median (Q2 or 50th percentile), the third-quartile value (Q3 or 75th percentile), and the maximum value. In other words, each five-number summary divides the data into quarters (each group contains approximately 25% of the data) and then uses these values to draw a boxplot; see the following figure.

To find the quartiles, the data are ordered, the ordered data is divided into four parts with approximately the same number of data points in each group. The quartiles are the cutoffs for separation of the data based on this division. That is, the first quartile (Q1) represents the value that approximately 25% of the data are below; the second quartile (also the median) represents the value that approximately 50% of the data are below; and the third quartile (Q3) represents the value that approximately 75% of the data are below.

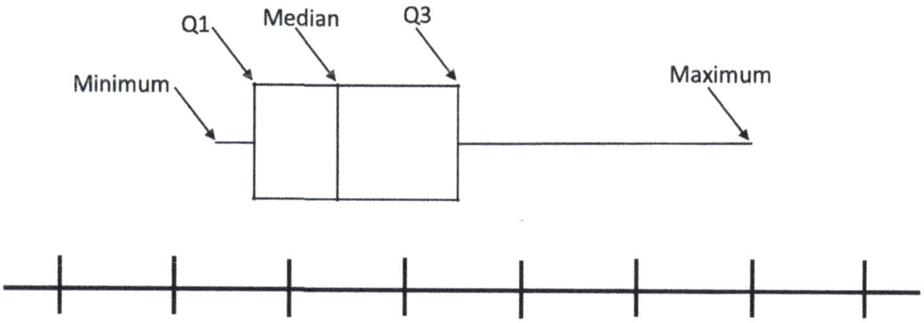

The following three boxplots visualize these measures of position for the three samples of students.

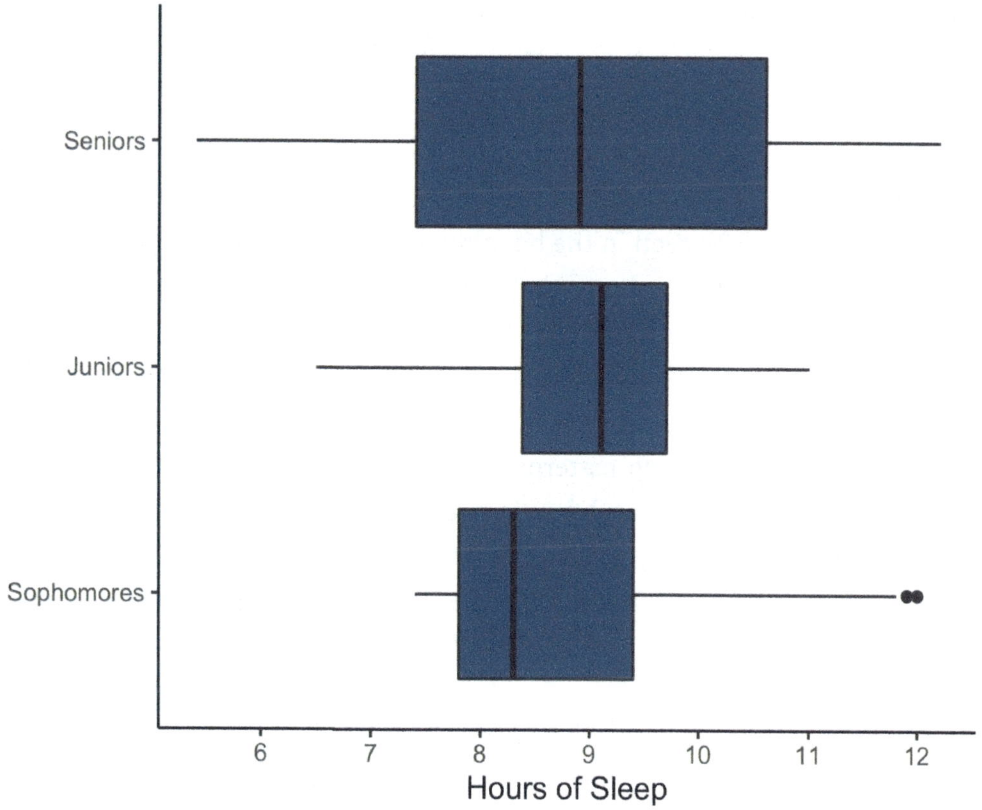

The vertical line in the middle of the boxes represents the median for that sample's values, the upper and lower edge of the box are drawn to the first- and third-quartile values, and the "whiskers" coming from the box are drawn to the minimum and maximum values. If the data contain an extreme value, also called **an outlier**, the whisker will stop at the largest or smallest data value that isn't an outlier. An **outlier** is defined as a value that is abnormally extreme. While the context of the data informs what should be considered an abnormally extreme value, a standard determinant used in practice is whether the point is more than 1.5 **interquartile range (IQR)** values away from the first or third quartile. The **IQR** measures the length of the box in the boxplot and can be computed by finding the difference between the first and third quartiles. This represents the range of the middle 50 percent of the data. For example, the boxplot describing the sleep data from the sophomores is built from the following five-number summary:

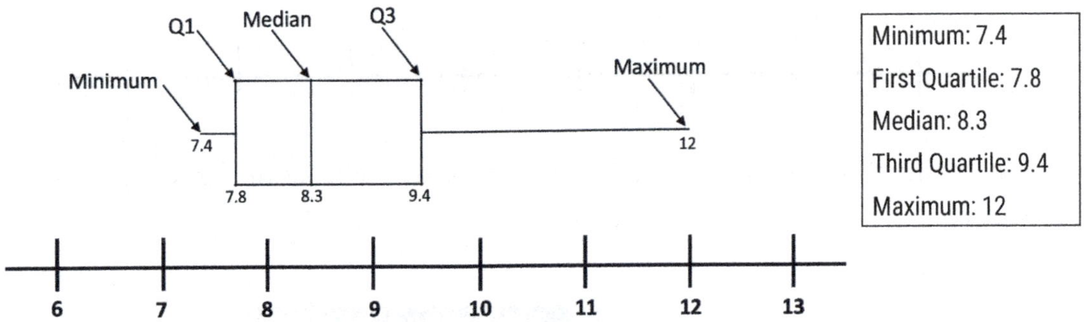

In this example, the IQR is equal to 9.4-7.8 = 1.6. An upper outlier would be a value that is greater than (1.5)*(1.6) + 9.4 = 11.8. This indicates that the sophomore data have four outliers—those can be seen in the boxplots as the dots past the end of the whisker, or in the dotplot. The dot plot shows that there are four such sleep times.

When comparing boxplots, we want to look for the "**overlap**" and "**separation**" among the different grades' sleep data. If the boxes overlap a lot, then the number of hours between each grade level's sleep patterns is not that different. If instead the boxes are clearly separated, such as one box having no overlap with another, then the sleep patterns are different. In this example, we see lots of overlap among the "boxes" in the graph, which means there is no clear difference in the sleep patterns of sophomore, junior, and senior students.

To compare the grade levels' variability, we can look at the length of the box, also known as the range of the middle half of the data, and observe that the juniors appear to have the smallest variation in sleep hours and that seniors have the largest variation. The sophomores have an IQR = 9.4-7.4 = 2, the juniors have an IQR = 1.35, and the seniors have an IQR = 3.2. We can visually see that the IQR for the seniors is the largest, noting that their box is the longest. Additionally, we can see that the seniors have the largest range of sleep hours, with a minimum value that is much smaller than the others. However, the median amount of hours of sleep for the three grades is very similar. The median lines are around 8–9 hours for all grade levels. The IQR is a natural measure of variability to use when choosing the median as a measure of center.

Dotplots, histograms, and boxplots all provide graphical visualizations of the distributions of sleep patterns for the sophomores, juniors, and seniors. These visualizations help us identify and compare patterns in the sleep times for the students in the different classes. They help us answer our investigative question: How do the sleep patterns of the students in the different grades compare?

In comparing the distributions of the three grade levels of students' sleep patterns, we see that the shapes of the distributions are vastly different. Whereas the seniors' sleep times have an approximately uniform distribution, the juniors's sleep patterns follow a mound-shaped, symmetric distribution, and the sophomores' sleep patterns have a skewed distribution. The seniors have the most variability in sleep, both because there are approximately the same number of seniors sleeping all the different hours of sleep, and because they have the largest range. However, the amount of sleep the sophomores typically get is concentrated around 8 hours of sleep and never less than 7 hours of sleep. In the case of the juniors, they have the least amount of variability in sleep, concentrated around 9 hours, and fewer and fewer people sleeping hours above or below 9. Although there appear to be differences in the variability of values for each of the grade-level samples, they all appear to have a similar number of typical hours of sleep, as visualized by the similar center value seen in all three of the plots. The dotplots and the histogram reveal typical sleep to be around 9 hours for the seniors and juniors and approximately 8.5 hours for the sophomores. The boxplots show that the medians are also close in value around 8 or 9.

It is important to note that all of the comparisons drawn at this point have been made by examining key features of the visualizations. Visualizing the distributions can provide an intuitive feel for the data and can also be used as a tool to see overall patterns and draw conclusions.

To further check our interpretations based on the visualizations, we will compare various summary statistics. Here is a table illustrating the mean, median, standard deviation, range, minimum, maximum, first-quartile, and third-quartile values for each of the grade levels.

Summary Statistics:

Column	*n*	Mean	Std. dev.	Median	Range	Min	Max	Q1	Q3
Sophomores	180	8.8	1.3	8.3	4.6	7.4	12	7.8	9.4
Juniors	180	8.9	1.0	9.1	4.5	6.5	11	8.35	9.7
Seniors	180	9.0	1.9	8.9	6.8	5.4	12.2	7.4	10.6

As we can see, all of the comparisons drawn from our visualizations have been verified. To draw meaningful comparisons between groups, we must consider all of the summary statistics. For example, both of the following histograms have the same means and medians (represented by the red and green lines), but their variability is drastically different. One distribution is very spread out, and the other is very concentrated around the mean and the median values.

Instead of relying only on summary statistics, students should be encouraged to visualize the distribution. Doing so enables one to see the numerical summaries, as well as the overall patterns in the data more clearly. In addition, a visualization can help one understand the interplay between measures of center, measures of variability, and the shape of the data.

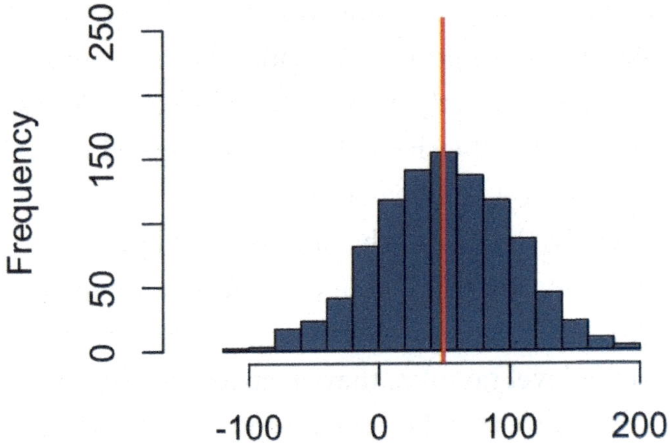

High Variability

Now, let's reflect on the interpretations we drew previously. Evidence in prior studies and literature implies that students who get appropriate amounts of sleep should be healthier and thus, implicitly, should be able to perform at higher levels. It is important to note the limitations of the data. The data were collected for one single night's sleep; therefore, all of our results are solely results about that Wednesday night. We

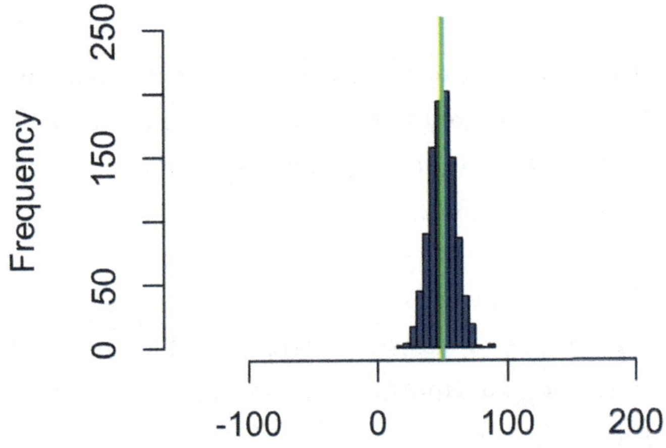

Low Variability

cannot adequately infer or generalize to other nights beyond our Wednesday night. While that is a limitation of this study, it is still worthwhile to compare the night's sleep of the high schoolers on the typical Wednesday night.

Comparing the samples from the different grade levels by looking only at the number of students who are getting enough sleep according to the recommendations, we see that about half of the seniors get the recommended amount of sleep compared with the others. Looking at the boxplot, we can see this by focusing on the top half of the box. The top half of the boxplot for the seniors goes from about 9 to 11 hours, while the top half of the boxplot for the juniors goes from 8.5 to 9.5 hours. Out of all the grades surveyed, the juniors recorded the single minimum number of hours slept.

On the lower end, no sophomores had less than 7.4 hours of sleep, which is interesting compared with the lowest junior value of 6.5 hours and the lowest senior value of 5.5 hours of sleep. Although the variation of sophomores' recorded number of sleeping hours is slightly greater than that of the juniors, the fact that the sophomores did not dip below 7.4 hours of sleep on a typical Wednesday night is encouraging, because NIH's recommendations state that people in this age bracket need 9–10 hours each day.

Considering that NIH provides 9–10 hours as the recommended amount of time for students of this age, the majority of the students are not achieving that goal in a typical night's sleep. However, the third quartile for seniors is a value above 10 hours, indicating that at least 25% of the seniors in the sample exceeded the recommended amount.

In the context of sleep, and because the typical amount of sleep is similar across the grades (the means and the medians for all three grades are very close in value), putting more weight on the variability in the values when drawing conclusions is justifiable. Along these lines, the juniors have the lowest variation. However, some of the juniors do get less sleep than some of the sophomores. Students can be encouraged to wonder and brainstorm as to what might explain the difference in variability among the three groups. We can summarize our comparisons in the following manner.

Seniors: The seniors meet the requirement of 10 hours of sleep most often. The seniors also had the most people furthest from the recommended range of 9–10 hours. The third-quartile value is above 10, indicating that at least 25% of the seniors sleep more than 10 hours.

Juniors: The juniors have the least amount of variability in values. They have the smallest standard deviation and range, indicating that the juniors have the most consistent amount of sleep for that Wednesday night.

Sophomores: The sophomores have the largest minimum value, indicating that no sophomores in the sample got below 7.4 hours of sleep, which at least meets the requirement for adults. The sophomores also have a high maximum value, indicating that several students get more than the recommended amount of sleep. This could possibly be due to them playing "catch-up" on sleep, depending on how many hours of sleep they had over previous nights. The sophomores also have a range of approximately 4 hours and a standard deviation of 1.2 hours, both indicating high variability, although not the highest for the three grades.

INVESTIGATION SUMMARY:

The main concepts developed in the sleep patterns investigation are:

1. We can visualize the distribution of a quantitative variable through a dotplot, histogram, and boxplot. Each of these graphical displays highlights different patterns in the distribution, which makes it important to look at all of them in order to make informed conclusions.
2. Distributions have different shapes. Distributions can be symmetric such as uniform and mound shaped or skewed such as having a long left tail or a long right tail.
3. Measures of center for quantitative variables are the mean and median. Measures of variability for quantitative variables are the standard deviation, the MAD, the IQR, and range.
4. We can use the shape of the distribution, the measures of center, the measures of variability, the maximum and minimum values, the quartile values, the identification of outliers, and the context of the problem to extract patterns from data and draw meaningful conclusions.

Follow-Up Questions

1. Provide an explanation or a definition of a distribution of a variable.
2. Sketch a distribution for which the mean is greater than the median.
3. Sketch a distribution for which the mean is equal to the median.
4. Sketch a distribution for which the mean is smaller than the median.
5. What do students gain from looking at different graphical displays of distributions of the same data? What specific features of comparison are facilitated by boxplots, histograms, and dotplots?

Investigation 1D.2: Restfulness

Goals of this investigation: Engage in the statistical investigative process and compare distributions for categorical data.

As noted in Investigation 1D.1, the National Institutes of Health recommends the following general amount of sleep for people of different ages:

Age	Recommended Amount of Sleep
Newborns	16–18 hours a day
Preschool-aged children	11–12 hours a day
School-aged children	At least 10 hours a day
Teens	9–10 hours a day
Adults (including the elderly)	7–8 hours a day

www.nhlbi.nih.gov/health/health-topics/topics/sdd/howmuch

Although the recommendations are specific, much variation exists related to the amount of sleep people get in reality. Some people claim they need less sleep than others, while others claim that they need much more. In some sense, each person has a baseline amount of sleep they believe they need in order to feel rested. In an effort to better understand how students within a school district are feeling when they come to school, teachers at the middle school and high school decide to conduct a small study to examine the sleep patterns of seventh grade and 11th grade students. The principals at the middle school and the high school selected 180 students from seventh grade and 11th grade and asked them to record how they felt on a typical Wednesday morning when they came to school. The students recorded whether they felt rested, somewhat rested, somewhat tired, or tired.

The data are recorded in the data set StudentRest.csv. Using the data in StudentRest.csv, answer the following investigative question:

How does the restfulness of the students in the different grades compare?

To answer the question, we need visualizations of the distributions of restfulness for the two student samples. Because the students recorded whether they were rested, somewhat rested, somewhat tired, or tired, each student can be identified with a category of restfulness; thus, the student rest data are categorical. Appropriate graphical displays to visualize the distributions of a categorical variable include pie charts and bar graphs. The displays for quantitative variables, such as dotplots or histograms, would not be appropriate for categorical variables because these visuals show sequential numerical relationships and categorical variables are not sequential between categories. Although pie charts and bar graphs visualize the same data, they may highlight different aspects of the data.

Let's begin by examining the distributions of the restfulness of the two samples of students through pie charts:

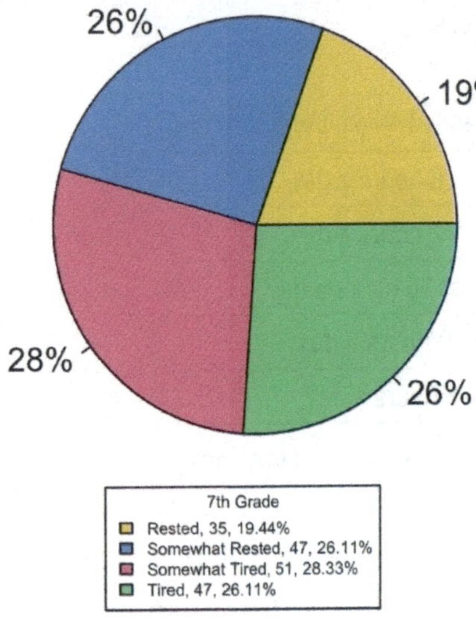

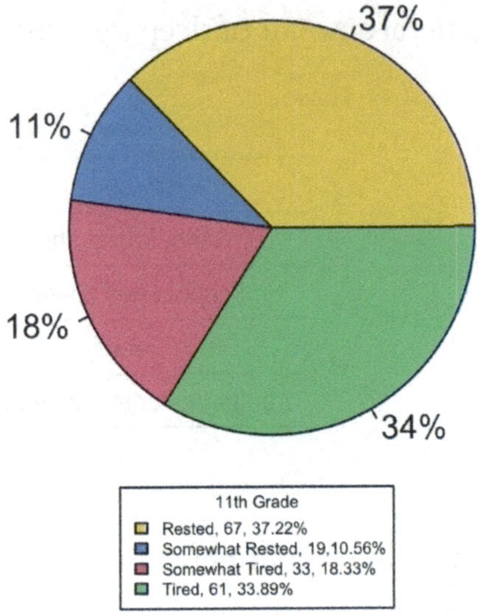

The pie charts show that the seventh graders have similar percentages of people across the categories (high variability), while the 11th graders are more concentrated in the rested and tired categories, indicating that the 11th graders have less variability. Bar graphs can also help visually compare the differences between the categories across the two grades.

The bar graphs reveal that the categories for the seventh graders have a similar number of responses, while the 11th graders favored two specific responses. This again shows that the 11th graders are more concentrated in two categories; thus, the 11th-grade sample has less variability than the seventh grade sample. Looking at the bar graphs, we can see that the modal

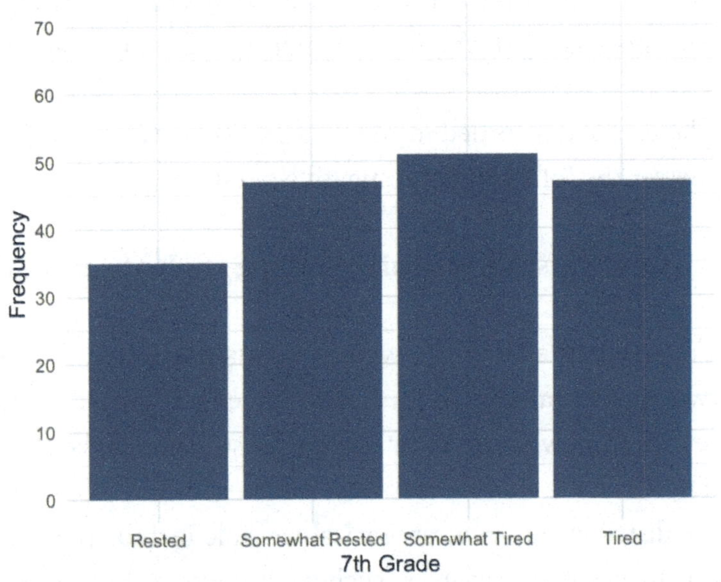

category for the seventh graders is category 3 (somewhat tired) and for the 11th graders it is category 1 (rested).

To further compare the samples of students' restfulness and see the exact counts in each category, we can visualize the data in a table and then examine the frequency and the percentage of data in each category for each grade.

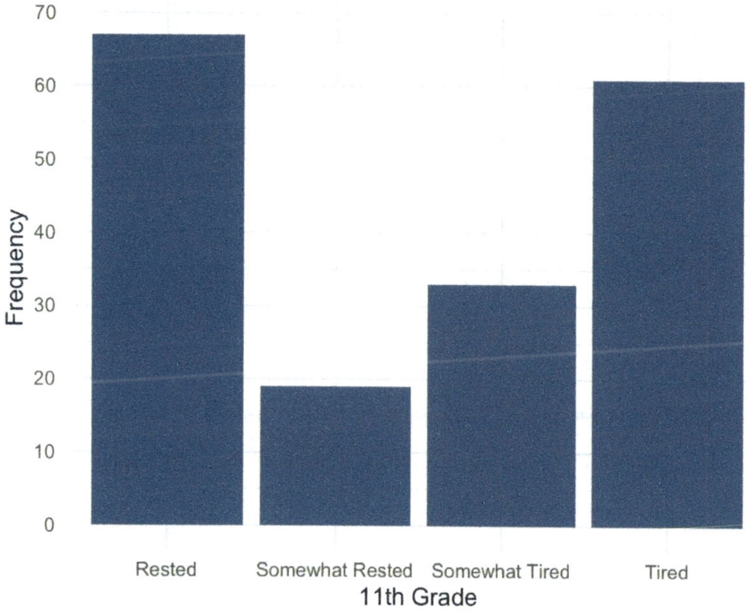

Frequency Table Results for 11th Graders:

11th Graders	Frequency	Percentage
Rested	67	37%
Somewhat rested	19	11%
Somewhat tired	33	18%
Tired	61	34%

Frequency Table Results for Seventh Graders:

Seventh Graders	Frequency	Percentage
Rested	35	20%
Somewhat rested	47	26%
Somewhat tired	51	28%
Tired	47	26%

The tables illustrate the percentage of responses in each category for each grade. We see that the seventh and 11th grade students exhibit different patterns of restfulness. Compared with the seventh graders, the 11th graders have a higher percentage of people on the two extreme ends. There is a 17% difference between the seventh and 11th graders in the Rested category and an 8% difference between the grades in the Tired category. Overall, it is harder to predict how rested a seventh grader will feel, because they are more variable regarding their levels of restfulness. The 11th graders are more predictable, because they are more likely to feel either rested or tired in the morning. The tables illustrate the percentage of responses in each category for each grade.[5]

5 Note that the table offers rounded up values whereas the graphs might not.

INVESTIGATION SUMMARY:

The main concepts developed in the restfulness investigation are:

1. We can visualize the distribution of a categorical variable through a pie chart, a bar chart, or a table. Each of these displays highlights different patterns in the data, so it is important to look at all of them.
2. To extract patterns from data and draw meaningful conclusions, we can use the modal category, the relative frequencies of the categories, and the graphical displays.
3. The variability of a categorical variable can be characterized as the amount of diversity or disagreement there is across categories. The more uniform across categories, the more variability.

Follow-Up Questions

1. How are graphical displays helpful for students to visualize distributions of categorical variables? What is more apparent to the reader from graphs than from tables?
2. Sketch the distribution of a categorical variable that is bimodal (i.e., having two modes).
3. Sketch the distribution of a categorical variable that is unimodal (i.e., having one mode).

UNIT 1E: Exploring Relationships between Variables

We are often interested in examining connections between variables. For example, we might want to know whether students' GPAs are linked to their SAT scores, whether eating breakfast is associated with better test performance, or whether liking to watch sports is linked to playing sports. In all these cases, we have two variables of interest, and we want to explore how the variables are related.

As we learned in previous units, there are two types of variables—categorical and quantitative. When we examine association between variables, there are three possible situations to examine:

- A categorical variable and a quantitative variable
- A categorical variable and a categorical variable
- A quantitative variable and a quantitative variable

An **association** is present if changes in one variable lead to systematic changes in the other. In this section, we use the words *association* and *relationship* interchangeably. In studies where we examine associations and relationships, we must have at least two variables. To study association and relationships, we need to identify the type of variables we have and understand which variable is the **response** variable and which is the **explanatory** variable for the posed investigative question[6]. A response variable is the variable with which comparisons are made. The **explanatory** variable is the variable that we hypothesize explains the outcome of the response variable. The response and the explanatory variable can be either categorical or quantitative. For example, if we consider the link between the number of cans of soda a person drinks a day and their glucose levels, we might be interested in answering the following investigative question: Is there an association between peoples' consumption of soda and their glucose levels? In this case, we might hypothesize that the more cans of soda one drinks, the

6 Sometimes the choice of explanatory and response variables are arbitrary. For a detailed discussion about the choices of variables and their impact on analysis see https://openintro-ims.netlify.app/

higher one's glucose levels. Therefore, the number of cans of soda would be the explanatory variable and the glucose levels would be the response variable. Both variables are quantitative.

Suppose we are interested in exploring how eating breakfast could be related to test performance. In such a scenario, we might set out to answer whether there is an association between breakfast eating and student test scores. Note that this investigation could be rephrased as a comparison question using the following investigative question: How do the test scores of students who eat breakfast compare with those of students who do not eat breakfast? In this investigation, we will use the association question.

Whether or not one eats breakfast is a categorical variable, and the student's test score is a quantitative variable. In this case, we might hypothesize that if one eats breakfast, then one might score better on a test. Therefore, we would say that breakfast eating is the explanatory variable and the test score would be the response. Similarly, we could examine links between the grade levels of students and their restfulness at school in the morning. In this case, both grade level and restfulness categories are categorical variables.

To examine whether values of one variable are more likely to occur with certain values of the other variable, we begin by making appropriate graphical displays that illustrate the likelihoods between the response and explanatory variables of interest.

The next three investigations examine relationships for the aforementioned three different combinations of categorical and quantitative variables. As we work through the investigations, it is important to consider the most appropriate graphical displays for visualizing the relationships. Questions about links and relationships between variables come up all the time in real-life scenarios. For example, people may be interested in the associations of particular diets with weight loss or gain, the impact of a specific treatment on cancer remission, the relationship between wealth and happiness, the association between gender and annual salary, or the relationship between education level and wealth. The list goes on. Because of this interest in drawing connections between variables, it is important to be able to examine links through data. In this unit, we will focus on noncausal relationships. In later units, we will discuss how associations between variables can be made causal (the explanatory variable somehow causes the outcome of the response variable).

Investigation 1E.1: Questions and Test Scores

Goals of this investigation: Illustrate how to visualize the relationship between a categorical variable and a quantitative variable.

A district has invested time and money in a professional development opportunity. As part of the professional development, each teacher participant must set a goal regarding a teaching strategy they plan on implementing in their classroom. Maria, one of the middle-school teachers in the district, is interested in understanding how students' use of questioning can help them increase achievement. Her goal for the professional development opportunity is to teach her students how to use questioning to guide their mathematical problem solving. Every time she gives an assignment or a test, she provides a space in the margin labeled "Guiding Questions" where students can jot down the questions they ask themselves while solving the problems. Maria believes that the students who make use of the questioning strategy will be more successful in solving the problems.

Her investigative question is:

Is there an association between posing questions and achievement? How do students' achievement for those who ask questions compare to students' achievement for those who do not ask questions?

To answer this question, Maria collected data from her class. She administered a Smarter Balanced Assessment Consortium (SBAC) Interim test and provided a side column for "Guiding Questions." She instructed students to use the space for noting questions that they thought of when solving the problems. Once she administered the exam, she collected the work and then recorded each student's score and whether they noted questions. The data are recorded in MariaTestData.csv. The data include several variables.

Maria notes that the observational unit in her study is the student. She has two variables of interest: (1) whether the student wrote relevant questions on the SBAC Interim test and (2) the student SBAC Interim test score. Whether the student asked questions is a categorical variable with two categories (yes or no). The student test score is quantitative.

To start her analysis, Maria checks how many students used the questioning strategy. Out of 35 students, she observes that 23 of them utilized the strategy she had been teaching, so approximately 66% of her students used noted questions during the exam.

Asked Questions	Frequency	Relative Frequency
No	12	0.34285714
Yes	23	0.65714286

Also, out of a possible 3000 points on the SBAC Interim test, the students' test scores had a median score of 2691 and a mean score of 2698. The distribution of the test scores is pictured in the following dotplot:

The students scored very well on the test, with the lowest score being around 2500 out of the 3000 possible points.

While the dotplot above displays the data for the students' tests scores, it does not help answer Maria's question about the association between posing questions and test scores. To examine the association between these variables, Maria needs to view the distribution of the test score broken down by whether the student noted questions or not. If she saw that the students who asked questions tended to score higher than those who did not, then she could say that there was an association between asking questions and the test score. If she saw that the scores of students who asked questions and those who did not ask questions had similar distributions, then she could say that there was not an association between asking questions and test scores.

To examine this, Maria can create two different histograms: one histogram representing the student test scores for those who asked questions, and another histogram representing the scores for those who did not ask questions. If there is an association between test performance and whether or not students asked questions, the histogram for the group who asked questions would be shifted further to the right (higher values) than that of the non-asking-questions group (lower values). The opposite could also be true; asking questions could also be connected to lower test scores. To easily see comparisons, we draw the histograms with the same scale. Here are the two histograms:

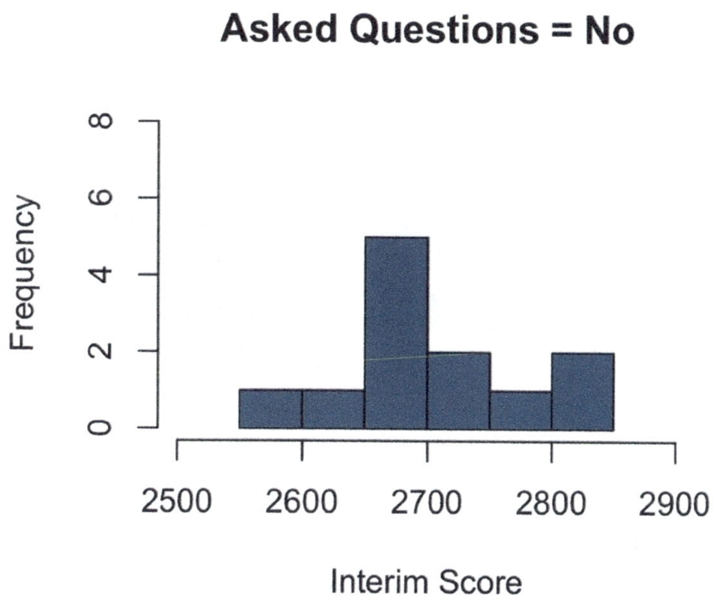

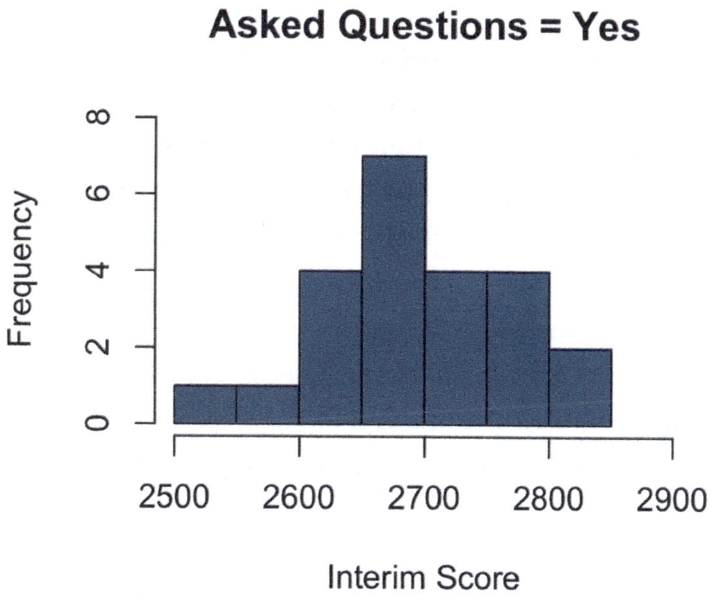

Using techniques discussed in the previous unit, we can see that the center of the distribution for the non-asking-questions group is similar to that of the asking questions group, around 2700. There is more variability in scores present in the question-asking group than the non-question-asking group. Also, because there is a difference in the amount of students in each group (12 people in the no question group and 23 in the question group), it is appropriate to use relative frequency histograms. While the histograms help us draw comparisons and view the association, as discussed in the prior unit, boxplots provide good visuals to compare distributions. As mentioned in previous units, when analyzing data, one does not need to only use one type of graphical display, because different displays allow one to make various observations that may be helpful in drawing conclusions.

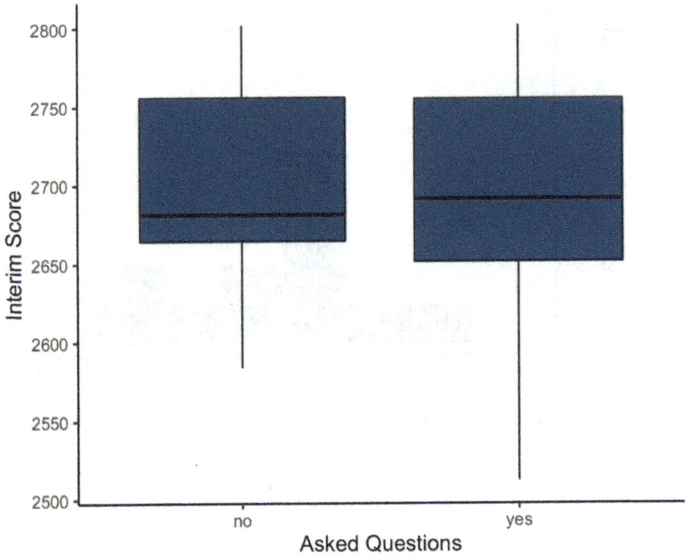

The boxplots illustrate that there is a large amount of overlap between the two groups' test score values and little separation between the test score values. The data indicate that questioning does not seem related to achievement on this test in any way. The IQRs of the test scores are similar across both groups of students—those who used questioning and those who did not. The middle 50% of the distributions have ranges of similar scores for both groups, as is the case for the fourth quarter of the distributions which display ranges for the higher scores. The major differences between the two distributions occurs in the first quarter where those asking questions had a much wider range of lower scores than those not asking questions. We can see this because both boxes have similar heights. In addition, their median score values are also similar, as depicted in the boxplots. However, the whisker of the question group reaches lower than that of the no-question group, indicating a wider range of scores for the question askers.

As a teacher, Maria could now use these analyses to inform her teaching. For example, it could be that the level of problems on the test were not the type that may be affected by questioning. Answering easy test questions correctly might not be affected by asking questions. On the other hand, questions that are complex, require multiple steps, and are geared toward problem solving could benefit from the strategy. This idea could prompt Maria to evaluate her tests to see whether the questions on the test varied in difficulty. In addition, the students in Maria's class scored very high on the Interim test, supporting the idea that the students found these problems easy. Maria could also undertake a more sophisticated study in which she examines the exam scores item by item. In such a study, it might be interesting for Maria to uncover which kinds of items benefited from using the questioning strategy.

> **INVESTIGATION SUMMARY:**
>
> **The main concepts developed in the questions and test scores investigation are:**
>
> 1. To visualize the association between a categorical variable and a quantitative variable, we can use the appropriate graphical displays for quantitative variables, broken down by the categories for the categorical variable. When given quantitative variables, dotplots, histograms, and boxplots can be separated by categories to compare the data.
> 2. Boxplots can easily show the overlap and separation of the quantitative variable by category.

Follow-Up Questions

1. Is there an association between the amount of time it takes a person to complete a sudoku puzzle in the morning and their feeling of tiredness when they wake up? Use the SudokuSleep.csv data to carry out this investigation.
2. Is there an association between achievement on the Levels of Conceptual Understanding in Statistics (LOCUS) exam and whether or not the test taker ate breakfast the morning of the test? Use the LOCUS.csv data to carry out this investigation.

Investigation 1E.2: Movie Budgets and Revenue

Goals of this investigation: Illustrate how to visualize the association between two quantitative variables.

We often hear that big blockbuster movies cost a lot of money to make. But do these big-budget movies pay off? Specifically, we are interested in understanding the answer to the following investigative question:

Is there an association between a movie's budget and the amount of money it makes worldwide?

To investigate this question, a data set was constructed for the large movies made between 1999 and 2018. A total of 5222 movies are included in the data set. The data can be found in Movies.csv[7].

The unit of observation in these data is a movie. For each movie, two quantitative variables are collected: (1) the budget spent on the movie in millions and (2) the worldwide gross revenue for the movie in millions. We hypothesize that as the budget increases, the revenue increases. In this sense, the budget is the explanatory variable and the revenue is the response. Because the explanatory variable is quantitative, it defines changes in values of the explanatory variable to be compared with respect to changes in the response variable. For example, as the budget increases, the revenue tends to increase as well. We can examine this change in association visually by looking at a scatterplot. A scatterplot is a two-dimensional dotplot. In previous units, we have used dotplots to visualize a single quantitative variable, but in this case, we have two quantitative variables. When we have two quantitative variables, the scatterplot is an appropriate data visualization.[8] It shows the values of both the explanatory variable (represented on the x-axis) and the response variable (represented on the y-axis) simultaneously by placing a dot for each state at the point on the coordinate plane that represents the two values. Essentially, it uses the data points as an ordered pair and plots the point with respect to the x- and y-axis.

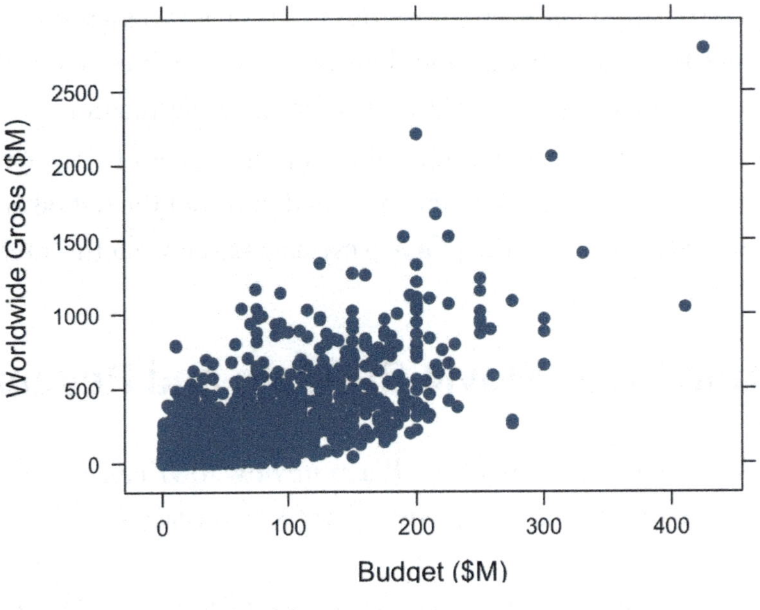

7 This data set was taken directly from StatCrunch's available data sets.

8 A line graph may be used to visualize a quantitative variable over time, where time is considered a quantitative variable as well. A line graph works well to visualize repeated measures of a quantitative variable over time (also referred to as time-series data). In general, however, when looking at two separate quantitative variables, scatterplots are an appropriate graphical display.

The scatterplot reveals that there is an upward trend—the higher the budget for a movie, the higher the worldwide revenue for the movie. There are a few movies that stick out on the scatterplot. For example, looking at the data set in Movies.csv, we can find the movie that corresponds to certain points. For example, *Avatar* is the movie in the far-right corner of the scatterplot—it had the largest budget and also the largest revenue. *Titanic*, on the other hand, had a much smaller budget than *Avatar*, approximately half, yet was the next-highest-grossing movie worldwide. *Pirates of the Caribbean* had a budget the size of Avatar, but had a revenue of only approximately 1000. In scatterplots with this many points, it is often difficult to see an overall trend merely because there are so many points on the graph. However, in this case, the upward trend is obvious, showing that budgets and revenue are positively associated.

To specify the association further, we can try to model it using an equation. As noted, the trend is upward sloping (positive association). The trend also appears to be somewhat linear. There is not a clear way to eyeball this; however, one possible way might be to draw an enclosure around the points and see that most of them fall in a narrow oval-like shape. In other words, the cloud of points follows a linear pattern. If it did not, we might see a U-shaped pattern (a cloud of points that looks like a parabola or quadratic), no pattern at all (a cloud of points has no pattern and is just scattered everywhere), or maybe an exponential pattern (a cloud of points that looks like an exponential function increasing on one side and tending toward the x-axis on the other). Because our pattern appears somewhat linear, we will try to model it with a linear function.

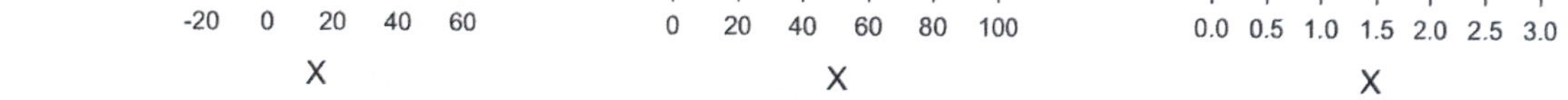

Quadratic Pattern, Random Pattern, Exponential Pattern

Specifying a linear model means writing an equation for a line that best summarizes the relationship between the variables. To write an equation of a line, we must specify two pieces of information: a slope and a y-intercept. To estimate what an appropriate slope and y-intercept might be for this scatterplot, we can informally fit a straight line for the scatterplot in such a way that we believe best captures the trend (this can also be done with a piece of spaghetti or a string).

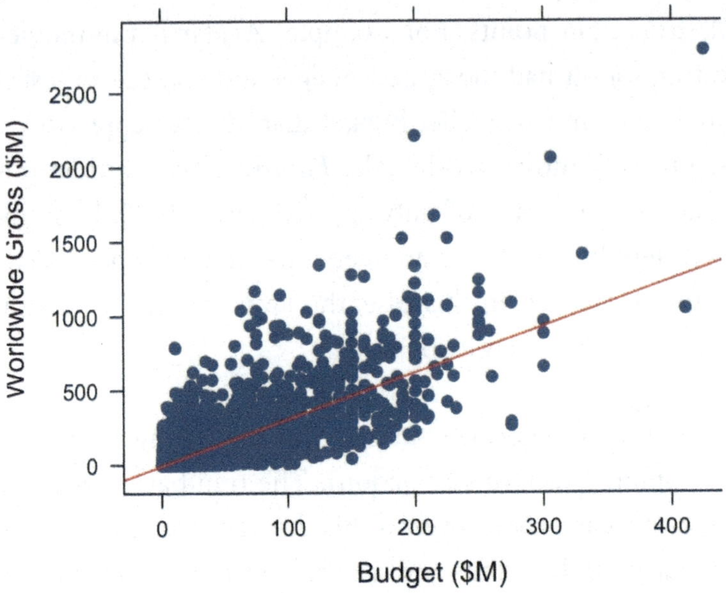

There are many ways to informally draw a line of best fit. One such example is drawing a line that captures the trend by minimizing the overall distance of the points to the line. This is difficult, if not impossible, to do mentally and visually; however we can get close. When asked to place a line of best fit, students may connect the left-most and the right-most points, or split the points in half to land above and below the line, or start the line from the origin (see Nagle, Casey, and Moore-Russo, 2017; Casey and Nagle, 2016; and Casey, 2016, for research on these types of student conceptions). We can use two of the points that landed close to the line to estimate the line's slope, (300, 800) and (190, 600). Using these points and the slope formula $(\frac{\textit{change in y}}{\textit{change in x}}) = \frac{y_2 - y_1}{x_2 - x_1} = \frac{\Delta y}{\Delta x}$, we estimate the slope to be around 200/110 = 1.8. If we continued to draw the line until it crossed the y-axis, we could estimate the y-intercept to be around 0. Therefore, we can model the movie revenue as a function of the movie budget with this line:

Worldwide Gross Revenue = 1.8*(Movie Budget) + 0

Where 1.8 is the slope and 0 is the y-intercept. Statisticians typically use the letter a to represent the y-intercept and the letter b to represent the slope. We can interpret the **slope** as the average predicted change in the response variable for a one-unit change in the explanatory variable. *This is interpreted as that on average, as the movie budget increases by 1 million (one unit of the explanatory variable), the worldwide revenue increases by 1.8 million.* The words *on average* are used in the interpretation because the equation is not deterministic. Instead, it is a predictive equation taking into account the variability of the data around the line. In this sense, the slope is indicative of the association present between the explanatory and response variable.

While the association found is predictive, it does not imply a causal relationship. In other words, we cannot state that the budget spending causes movie revenue to increase or decrease. While interpreting the slope and making predictive statements, it is also important to understand that we are not implying that the movie budget is the only variable that might explain movie revenue worldwide.

> **INVESTIGATION SUMMARY:**
> **The main concepts developed in the movie budgets and revenue investigation are:**
> 1. To visualize the association between two quantitative variables, we can use scatterplots.
> 2. From the scatterplot, we can visualize the type of association that might be present between the variables and estimate an equation to model this relationship.
> 3. If the association between two quantitative variables looks linear, we can model it by finding the equation of a line that best represents the data set.
> 4. The slope of the linear equation tells you how a one-unit change in the explanatory variable will predict the change in the response variable.

Follow-Up Questions

1. Is there an association between student SAT scores and student college GPA? Use GPADataSet_1000.csv to conduct this investigation

Investigation 1E.3: Body Image

Goals of this investigation: Illustrate how to visualize the association between two categorical variables.

Many Americans struggle with their weight. According to the Centers for Disease Control (CDC), more than a third of Americans are obese (www.cdc.gov/obesity/data/adult.html). While many struggle with weight issues, Americans are also constantly inundated with images and pressures to obtain a perfect body. Women are pressured as the media inundate them with pictures of celebrities quickly recovering from having babies, and of models on the cover of magazines. References to women's appearances occur more frequently than discussions about accomplishments. Men also feel body image pressure (see for example, www.theatlantic.com/health/archive/2014/03/body-image-pressure-increasingly-affects-boys/283897/). They are increasingly worried about gaining more muscle and being thin. Considering these societal norms, a health class at a high school is interested in understanding if males and females have different feelings about their body image. They pose the following investigative question:

Is there an association between gender and the way they feel about their weight?

To answer this question, the health class used an existing data set collected through a survey of 236 undergraduate students at a major university9. The survey asked students to report their gender and their feelings about their weight. The survey question that asked about weight was "Do you feel that you are underweight, about right, or overweight?" The data are presented in BodyImage.csv. A total of 229 students (out of the 236) answered the survey questions regarding their gender and their feelings about weight. The following contingency table shows the breakdown of students' opinions about their weight by their gender. A **two-way table** (often also referred to as a contingency table) is a table that displays the distribution of one categorical variable in the columns and one categorical variable in the rows. Each cell of the contingency table displays the number of people that fall in the row and column category.

	About Right	Overweight	Underweight	*Total*
Female	107	32	6	145
Male	56	15	13	84
Total	163	47	19	229

We could also include the **joint relative frequencies** for each category in the table. Of the total respondents, 47% are females who believe they are about the right weight; 14% are females who believe they are overweight; and only 3% are females who are underweight. The joint relative frequencies are also given for the males. We consider the male/female variable the explanatory variable and the opinion on weight the response variable.

The table also shows the **marginal relative frequencies**. These come from the total columns. We see that the marginal relative frequency of the about right category is 71%

9 This data set is a real data set collected and then posted on StatCrunch for others to use.

of the total respondents; 21% chose the overweight category, and only 8% chose underweight. The marginal relative frequencies also reveal that we had 63% of females respond to the survey, as opposed to only 37% of males.

	About Right	Overweight	Underweight	*Total*
Female	47%	14%	3%	63%
Male	24%	6%	6%	37%
Total	71%	21%	8%	100%

Because of this imbalance of totals between the genders, it is important to look at the relative frequencies in each of the weight categories within each gender. We thus examine the **conditional relative frequencies**. For example, the conditional relative frequency of about right given someone is female is $107/145 = 0.73$, and the conditional relative frequency of overweight given someone if female is $32/145 = 0.22$.

Because we are interested in comparing how the different genders answer the weight question, we condition on gender. In other words, we find the relative frequencies within a gender. For example, looking solely at the female responses, we see that a majority of the 145 females believe they are just right, some believe they are overweight, and a few believe they are underweight. For the males, we see a similar pattern, but the number of males who believe they are overweight and underweight (15 and 13, respectively) are closer in value to each other than that of the females in those two categories (32 and 6, respectively). As mentioned, because there are not the same number of females and males surveyed, we should not compare frequencies in the categories across genders. Instead, we compare the conditional relative frequencies in each category, which would provide a better picture of the differences between the two genders' feelings about weight.

	About Right	Overweight	Underweight	*Total*
Female	107 (73.79%)	32 (22.07%)	6 (4.14%)	145 (100%)
Male	56 (66.67%)	15 (17.86%)	13 (15.48%)	84 (100%)
Total	163 (71.18%)	47 (20.52%)	19 (8.30%)	229 (100%)

The conditional relative frequencies in parentheses show the row percentage for that particular category. For example, approximately 74% of female respondents and 67% of male respondents are in the about right category. Approximately 22% of female respondents and 18% of male respondents are in the overweight category. Looking at the conditional relative frequencies, we see a large difference between males and females in the underweight category. Of the males participating in this survey, approximately 15% believed they were underweight, compared with only 4% of females.

In addition to the two-way table, we can visualize the different categories of weight feelings across genders using pie charts and bar graphs. As discussed in prior units, these two graphs are best used to visualize categorical variables. In this case, because we have two categorical variables, we will be creating two pie charts or two bar graphs, one for males and one for females, and then we will compare the distribution of feelings about weight across the categories.

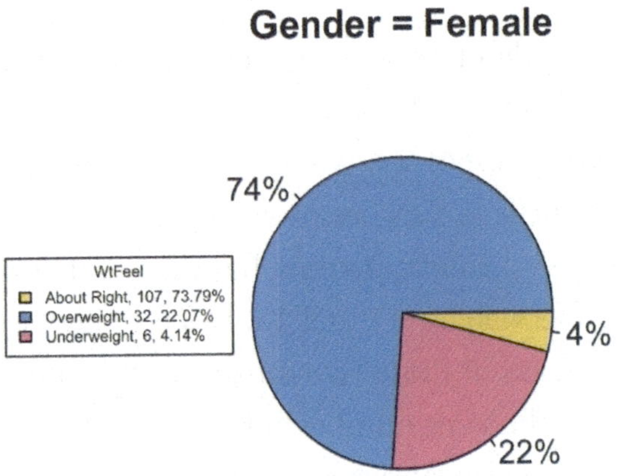

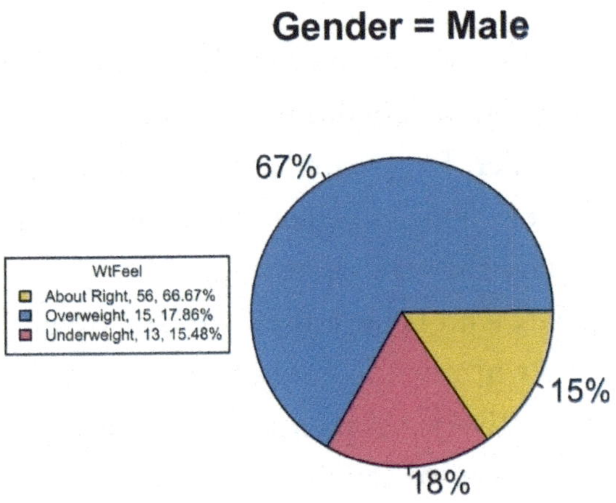

The pie charts show that the about right category dominates for both genders. It also shows that there is a difference in the underweight category for males and females. For the 229 people who answered the weight question, more males believe they are underweight than females. It is hard to make a visual comparison for the overweight category across the genders because they appear to be similar, so we will consider a bar graph in which the vertical axis represents the relative frequencies and thus could provide a clearer visualization for the comparison. As explained previously, we use the relative frequencies rather than the frequencies because the relative frequencies allow us to compare groups of different sizes. In this case, because there are more females included in the data set, it is important to use the relative frequency to compare the distributions.

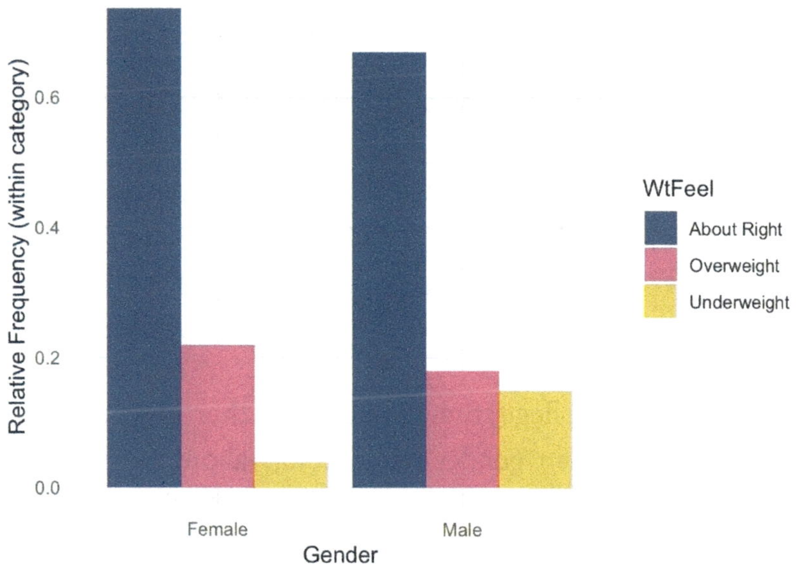

This bar graph depicts the distribution of weight feelings by gender and shows the differences in feelings about weight with possibly more clarity than the pie charts. This is because it is easier to compare the heights of each bar by color across the two categories.

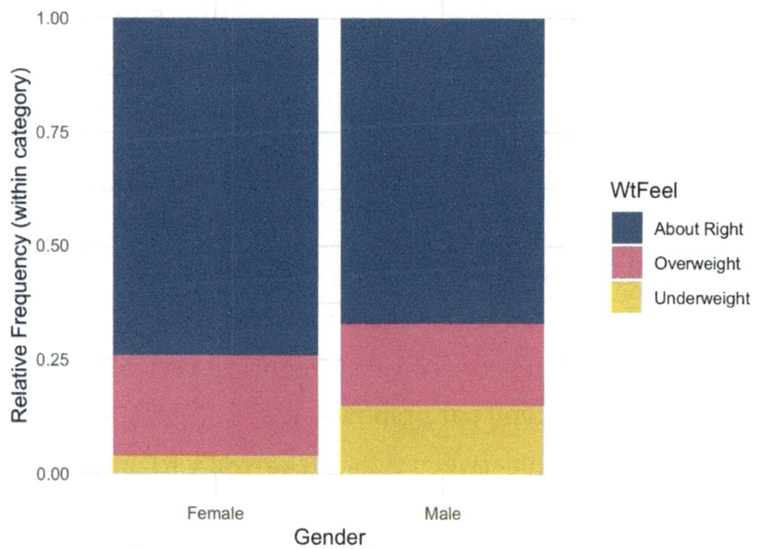

In addition, we add a stacked bar graph, which makes it easier to compare the distributions of how males and females answered the weight question. We see that the relative frequency of males who answered underweight (yellow) is larger than the relative frequency of the same category within females. We also see that the about right category (blue) was relatively more popular within females than males. It appears that the overweight relative frequency is about the same across the two genders. The stacked bar graph can help us see comparisons between the relative frequencies of many categories across the conditioning category (in this case, gender).

From the stacked bar graph, we see that there are differences between males and females in each category, but the relative frequency differences are slight for the about right category and the overweight category. However, the relative frequency differences are large for the underweight category.

Overall, in this group of 229 college students surveyed, there does appear to be a potentially weak association between gender and feelings about weight. This association is most evident in the difference between underweight feelings. More males tend to feel that they are underweight compared with females. If there had been no association between gender and feelings about weight, we would have seen approximately the same distribution of feelings about weight for males and females. But because we see a difference in the underweight category, we can say that a potential overall association is present. However, because only 8% of the entire sample falls into the underweight category, someone might claim that there is no association because the distribution is similar for 92% of the sample. This is a valid conclusion as well. At this point, a more formal statistical test that detects whether the association is present or not would be in order. Such a test is called a chi-squared test for independence and is discussed in the context in inference. For this investigation, informal conclusions that are in some way supported by the data are welcomed.

INVESTIGATION SUMMARY:

The main concepts developed in the body image and gender investigation are:

1. To visualize the association between two categorical variables, we can use contingency tables, pie charts, and bar graphs.
2. To examine the association between two categorical variables, we consider the relative percentages within each category we are comparing. This is because there might not be the same number of observational units sampled in each of the comparison categories.

Follow-Up Questions

1. Is there an association between people's belief about whether there is an appropriate emphasis in colleges and universities on college sports and people's beliefs about the kind of impact college sports have on academics at a college or university? To help answer this question, a survey of 2918 people was conducted. In the survey, people were asked whether they believed that colleges and universities put too much emphasis on sports, and they were asked whether they believed that having sports at a college or university had a negative impact on the academic experience. The survey responses are collected in ResponsestoCollegeSportsSurvey.csv.

References for This Unit

Casey, S. 2016. Finding what fits. *Mathematics Teaching in the Middle School* 21(8). www.nctm.org/Publications/Mathematics-Teaching-in-Middle-School/2016/Vol21/Issue8/Finding-What-Fits.

Casey, S., and C. Nagle. 2016. Students' use of slope conceptualizations when reasoning about the line of best fit. *Educational Studies in Mathematics* 92:163–77. https:/link.springer.com/article/10.1007/s10649-015-9679-y.

Nagle, C., Casey, S., and D. Moore-Russell. 2017. Slope and line of best fit: A transfer of knowledge case study. *School Science and Mathematics* 117: 13–26. https://doi.org/10.1111/ssm.12203.

UNIT 2: Toward Data Science

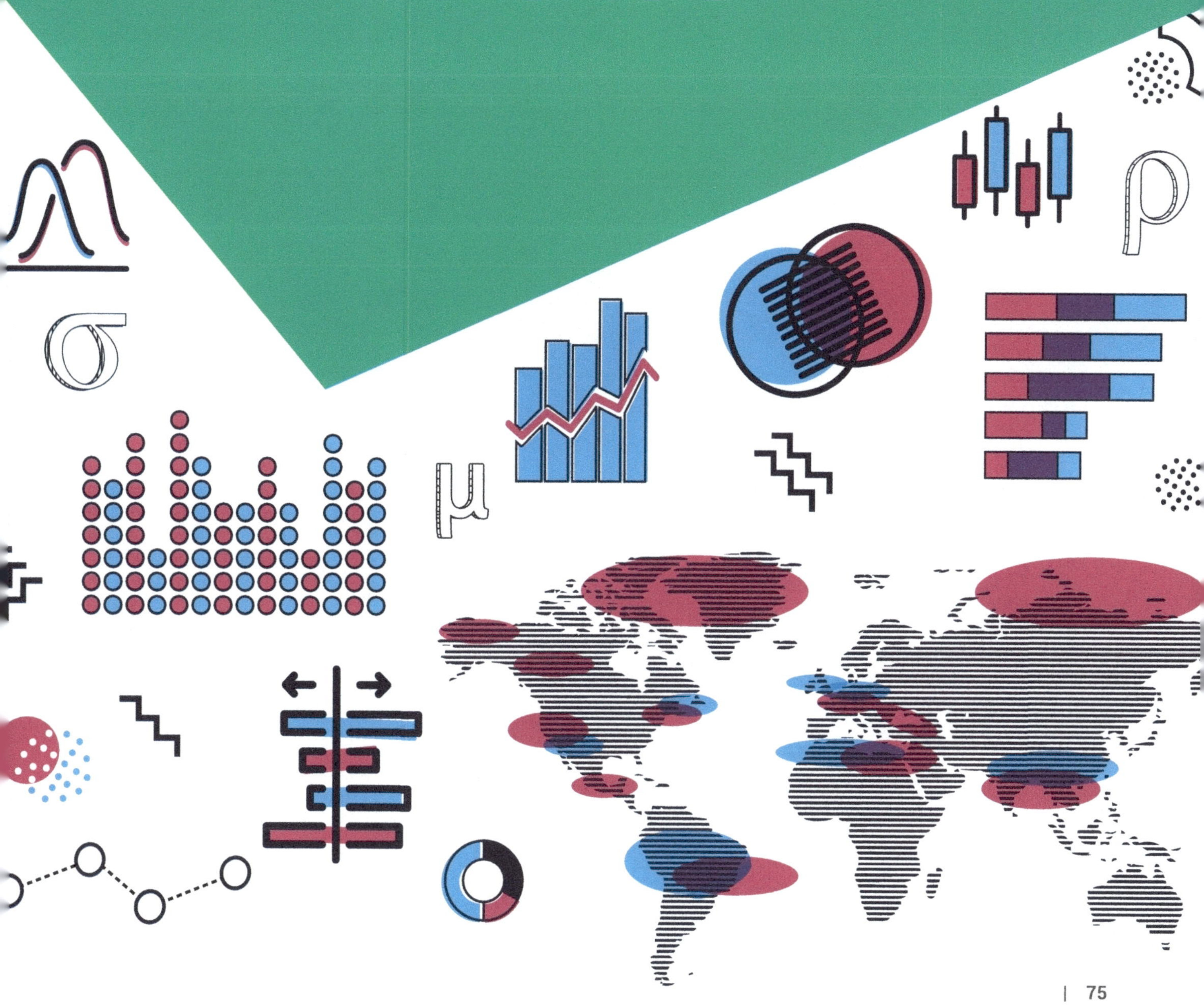

UNIT 2A:
Data in Our Daily Lives

Now more than ever, data are part of our daily lives. We collect data, we are inundated with data, we are shown data in myriad displays, and we are asked to interpret data to help us make decisions and form opinions *every single day.* Students are exposed to data in the form of text messages, pictures, sounds, and tweets through social media outlets and technological devices. There are also large amounts of data being collected daily and automatically, based on our behaviors. For example, data collected through an exercise-tracking device or purchase-history data from Amazon document daily routines in our life. Data may be produced by social networking (such as Twitter, Facebook, or LinkedIn) or gaming devices and smartphones, or streamed from satellites used to understand climate change. All of these examples of data fall under the general heading of "Big Data." The term originally referred to data sets of great size that had volume, variety, velocity, and veracity (Díaz, 2020); however, over time, it has expanded to include data that merely have characteristics that can potentially lead to great size. Big Data may include images, locations, and dates. These data are rich and worthy of analysis. We will refer to all of these types of data as "nontraditional data."

The focus of this unit is to show how nontraditional types of data can be collected, accessed, and analyzed in the elementary-, middle-, and high-school levels. If our goal as educators is to have students graduate high school statistically literate, we must incorporate curricula that address how to manage and analyze nontraditional data.

In this unit, we will introduce several investigations that deal with different types of nontraditional data. The sophistication and complexity of the investigations increase as the unit moves from Levels A, to B, to C. The investigations in the unit are not meant to provide an exhaustive list of types of nontraditional data to be used in the school curriculum at the different levels, but instead are meant to provide examples of interesting and relevant ways that nontraditional data can be introduced and analyzed. In Unit 1, we discussed the importance of questioning in undertaking the statistical problem-solving process. Posing questions can lead to data exploration and uncovering patterns within data. When dealing with nontraditional data, the role of questioning becomes crucial to carrying out worthwhile analyses and drawing appropriate conclusions.

Case Study 3: Graphical Displays

Graphical displays of data may provide powerful visuals for uncovering patterns in data and can help shed light on characteristics of the data distribution. Visualizations are an integral part of data analysis today. Standard graphical displays include bar graphs, pie graphs, frequency tables, and contingency tables[1] for representing the distribution of categorical variables; dotplots, scatterplots, histograms, and boxplots can represent the distribution of quantitative variables However, a graphical display can be *any* picture that provides visual information about important features of how the data are distributed. Typically, the K–12 grades focus on creating visuals for one or two variables at a time. However, with myriad data now available, thinking about multiple variables at a time and visualizing several variables within a data set in the same graphic is an important skill to develop. In fact, our interactions with data today are often through nontraditional graphics and interactive visuals that show data on many variables simultaneously.

For example, word clouds are now found in many news articles to visually represent text data. The size (magnitude) of a word reflects the frequency with which the word was used; higher-frequency words are larger. Word clouds are relatively mainstream data visualizations that essentially were not used in the early 2000s and are not included in current standards. Would it be possible to represent the frequency of text words in a bar graph? What might be some advantages of using a word cloud over a bar graph for these types of data? Or of using a bar graph over a word cloud?

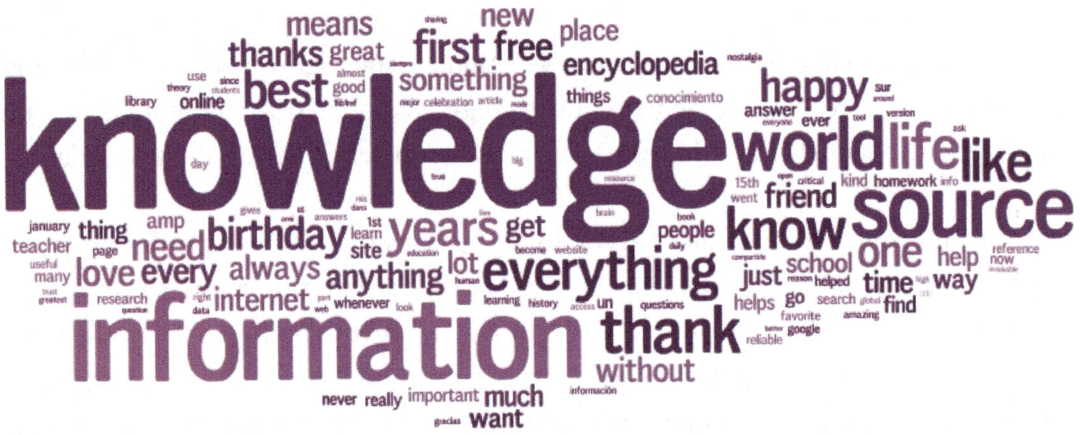

Graphic taken from https://commons.wikimedia.org/wiki/File:Wikipedia_15_-_Twitter_wordcloud.png. Original created by JSutherland (WMF) under the Creative Commons license

1 Frequency tables and contingency tables are also referred to as tabular summaries of data.

Another example of a graphic now used in popular media that is not taught in the current school curriculum is a map. Consider this graphic from *The New York Times*[2]. This choropleth map shows the coronavirus hot spots as of October 2, 2020. The color-coding shows a gradation of case severity in different areas of the United States. Would a boxplot be able to convey all of the information presented in this choropleth map? What are some advantages of using this map as a way to show patterns in the data?

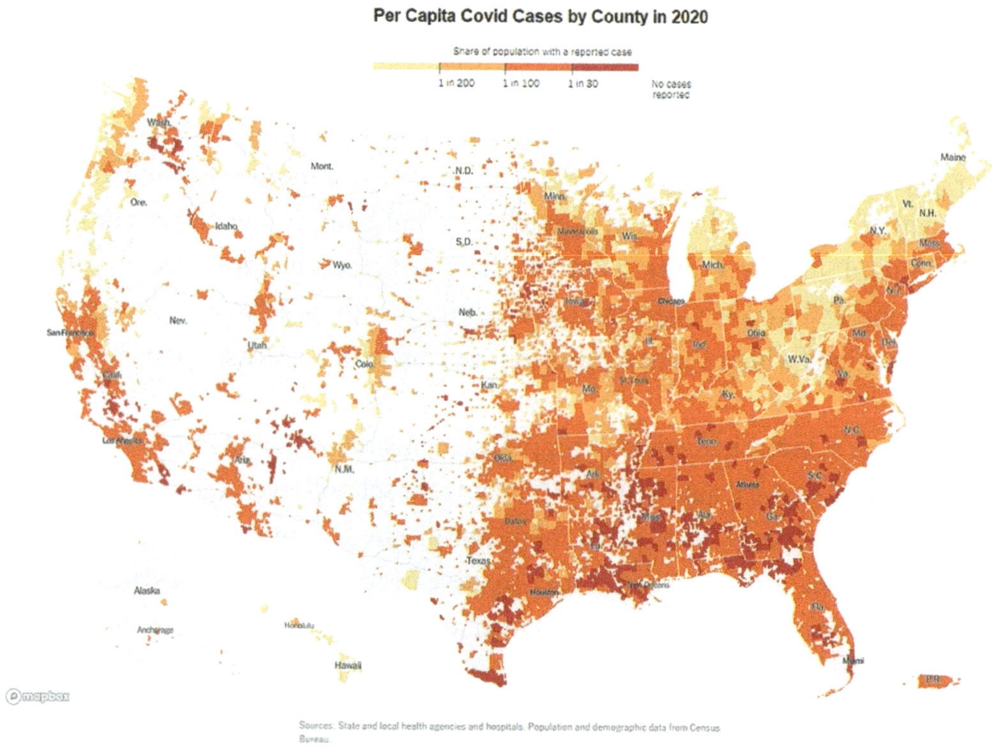

As specific investigative questions are posed, we can strategize about how to visualize the data in a way that provides a better answer to the investigative questions posed. Thinking about ways to visualize data on multiple variables is not a static and constricting process; instead, it can be quite creative. We want teachers and students to think "What would be the best picture to make sense of these data?" instead of automatically making a bar graph or histogram without considering the overall purpose. This helps students think about multiple variables at a time. Valuable discussions surrounding graphics and what students may wonder or notice about graphics are a focus of the collaboration between the American Statistical Association (ASA) and *The New York Times* called What's Going On in This Graph? (www.nytimes.com/column/whats-going-on-in-this-graph). As part of the *NYT*'s Learning Network, visualizations used in articles in the *NYT* are discussed live with students. Statisticians from the ASA moderate each

2 Graphic taken from www.nytimes.com/interactive/2020/us/coronavirus-us-cases.html?action=click&module=RelatedLinks&pgtype=Article. Used with permission.

week. This collaboration offers models of how to discuss rich data visualizations actually shown in the news with K–12 students. These types of discussions push students and teachers to think about the features of the data (multiple dimensions of the data) and how these dimensions can be represented in a meaningful way. By thinking critically about data visualizations, students are forced to think about all the different dimensions they want to represent.

Of course, there are good and not so good visualizations (see discussions about features of visualizations in Börner, Bueckle, and Ginda (2019) and King et. al (2021)). A good visualization presents the data in a way that highlights specific distribution characteristics and clarifying patterns present in the data. For the word cloud, the size of the words reveals the frequency of the words in a text, immediately highlighting the pattern of important words being used. Just by a quick glance at the word cloud pictured previously, we see that *information* and *knowledge* were the most frequently used words. For the choropleth map, the regions where the coronavirus is pervasive are easily spotted.

Being effective at summarizing data in graphical forms requires a high level of multivariate thinking, a skill that we aim to develop in current school-level standards and beyond. This unit will demonstrate how questioning can dictate the appropriateness of the type of display. It will focus on the dynamic and creative process of creating a data display that can capture the information in an innovative manner.

CASE STUDY SUMMARY:

The main concepts developed in the graphical displays case study are:

1. Distributions of variables can be displayed in many ways in the news.
2. Color and size can be used to display multiple variables in one graphic.
3. Interpreting graphical displays necessitates identifying all of the variables displayed in the graphic and how they are related to one another.

Next, we present another case study, followed by two investigations, that capture the creativity and subtleties that are important to master when learning to make sense of data in our daily lives. They also show how we can creatively represent the types of information we record while performing a data collection process. The case studies and investigations in this unit are meant to be introductory and thus are appropriate for early school levels, although exercises presented in this initial unit are also worthwhile for older students. The next sections, Units 2B and 2C, show how to extend the ideas to the secondary grades and illustrate more complex investigations of the same spirit.

A main purpose of Unit 2A is to understand that there are multiple ways to provide visual representations of data other than traditional displays. We encourage individuals to explore the creation of *different* visualizations for data that are not programmed into software. We promote this as a first step before individuals use dynamic statistical software. These skills require individuals to grapple with the variability in their data, examine the type of variables they have, understand the cases they have and what data are available, and represent multiple variables simultaneously. All of this forces individuals to make sense of their data.

Case Study 4: *Dear Data*

In 2016, information designers Giorgia Lupi and Stefanie Posavec published the book *Dear Data* (www.dear-data.com/theproject). This book chronicled a project they had carried out together for a year.

Used with permission.

Every week for one year, the two designers decided to choose a topic and collect data on that topic. At the end of the week, they would mail a postcard to each other that displayed their data in some type of graphical display that they drew by hand; on the back of the postcard, they included a key to their visualization.

One of the designers lived in Europe and the other lived in the United States. Topics included a week of positive feelings, a week of friends, a week of doors, and a week of drinks. Each designer could choose what data to collect and how to represent the data in any way she wanted. For example, for the week of doors, they tracked the doors that they passed through for the entire week and represented their data in the following ways:

Giorgia

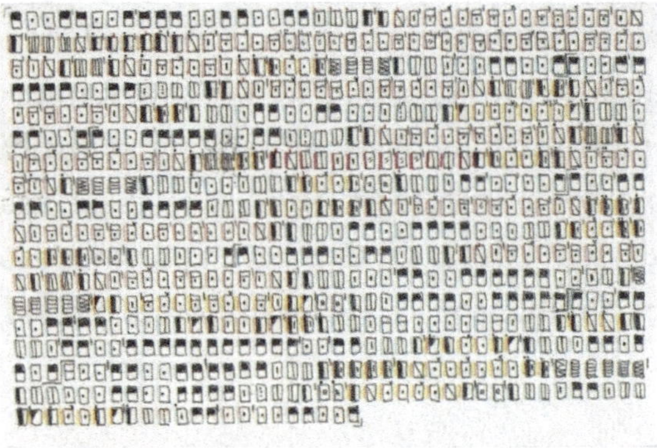

a week of doors

Stefanie

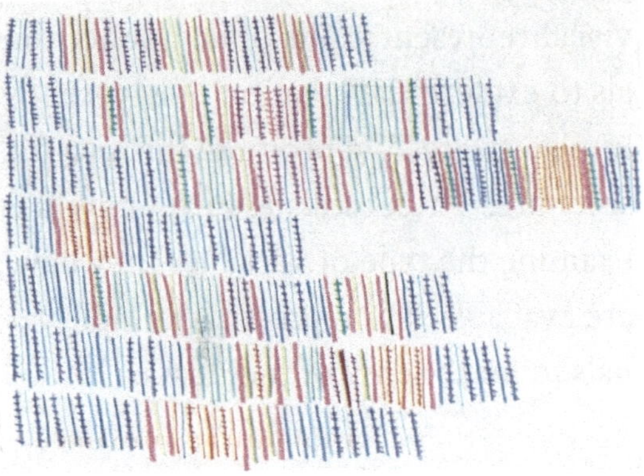

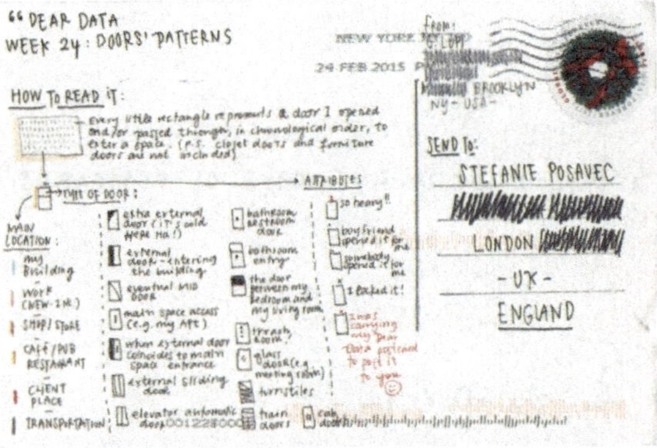

After spending more than six hours drawing this hyper-detailed card, Giorgia texted Stefanie as she posted it: "You need to know that if this one doesn't get to you, I won't redraw it. You'll see what I mean."

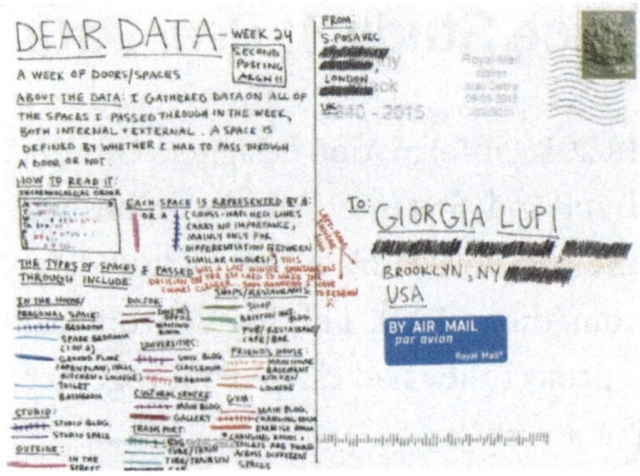

Unfortunately, while Giorgia's postcard arrived, Stefanie's postcard didn't, so she had to draw hers again (luckily is wasn't as detailed, but it was still supremely annoying).

Used with permission.

In Giorgia's image, we immediately notice that she represented the door as a rectangular shape, whereas Stefanie represented each door with a line. Giorgia represented the different kinds of doors by drawing characteristics on them. For example, the door between her bedroom and living room was represented with the top half of the door being colored black. We can see, even with only a quick glance at her display, that this door was one that Giorgia passed through often during the week. Color was introduced to represent the location of the door. In Stefanie's case, her data were visualized with specific types of lines representing each type of door. She also represented the doors in chronological order, with each row representing a day of the week. From this, we can immediately see that Stefanie went through many doors on the third day of the week.

The choice of topic for the week can be rephrased as an investigative question. For example, the data from the week of doors can help answer the following investigative question:

In what way do we interact with doors in our daily lives?

Giorgia and Stefanie then posed data collection questions to guide their data collection, such as:

- What type of door am I passing through?
- What is a distinguishing feature of the door I am passing through?
- What is the location of the door I am passing through?

We can see that on their postcards, each of these questions dictated a characteristic that they captured about each door. We call these characteristics the **variables** of interest. Each door was a **case** in their data sets (a case may also be referred to as a unit, an individual, or an observation in statistics). And for each case, they collected three measurements dictated by the data collection questions and variables of interest. Their representations displayed each case, and for each case, they displayed, in different ways, the three variables.

This required the authors to think about multiple variables simultaneously when they were designing the display for the three dimensions (variables) for each case. The requirement of drawing the data visualization by hand ensures that the authors could not take shortcuts in thinking about the meaning of their representation. Making the visualization by hand required deciding what things they wanted to be precise about and how that precision should be drawn.

This case study illustrates not only how data about our daily lives can be visualized in innovative and interesting ways, but also how data visualizations can reveal patterns in the data. We can see that both women interact with doors often in their daily lives. Their lives revolve around several locations, such as home, work, transportation, and outings. They are often crossing doors to go outside and inside, but more frequently the doors they cross are interior. This reveals that they might get to a certain place and stay there more than visiting many different places throughout the day. The data also reveal the regularity with which some doors are present in their lives, implying that there are specific spaces where they probably spend most of their time.

As the authors of *Dear Data* reflected on their project, they noted that data should be seen as a way to see the world around us that provides a starting point for discussions and not definitive answers to our questions. Giorgia discusses the project further in a TED Talk about how data humanizes us and reveals important stories about ourselves. The discussion, titled "How can we find ourselves in data?," can be found here: www.ted.com/talks/giorgia_lupi_how_we_can_find_ourselves_in_data.

CASE STUDY SUMMARY:

The main concepts developed in the *Dear Data* case study are:

1. Data can be displayed in many ways, not merely traditional displays that might be included in curriculum standards.
2. Creativity is often important in finding the best ways to spot patterns in multivariate data.
3. Drawing out data displays by hand is important because it encourages one to grapple with ideas of precision, modeling, and appropriate representation of variables measured in each case.

Building on the *Dear Data* case study, the following investigation illustrates the process of carrying out one week of *Dear Data* between an adult and an elementary-school student.

Investigation 2A.1: *Dear Data:* My Week of Happiness

Goals of this investigation: Work with real data we come across in our daily lives, work with multidimensional data, make data visualizations, and use questioning to gain insight about data.

Over the course of a week, data were gathered to answer the following investigative question:

What makes us happy?

A second grade child, Siena, and co-author Anna engaged in a seven-day data collection process where every day they recorded the instances, people, things, etc. that made them happy. The data collection questions were:

- What made you happy?
- What time of day were you happy?
- Why were you happy?

These questions were posed by Siena. They also recorded the day of the week. Therefore, they recorded a total of four variables (what, time, why, day of week).

To record their data throughout the day, Anna and Siena used 3x5 cards that could easily fit in their pockets, so they could pull them out when a new instance occurred and data needed to be recorded. The following represents the image of Siena's card for Monday:

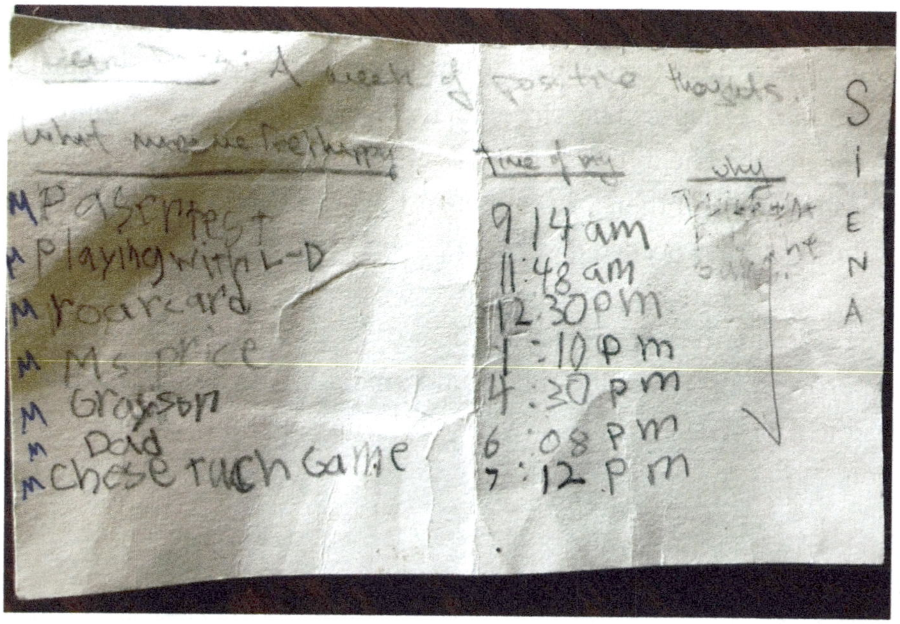

At the end of the week, both Siena and Anna took their data collected throughout the week and constructed their visualization.

Anna made the following visualization:

In this visualization, the rays from the sun/moon in the middle of the paper represented the 24 hours in a day. The days were represented on each hour as a little line (seven lines per day). The moon side of the graph represented the p.m. hours and the sun side represented the a.m. hours. The key at the bottom illustrates the things and people that made Anna happy, which were represented by different shapes. Further classification of why these things made Anna happy were given by the smaller details added to each of the shapes. Therefore, each case of happiness was represented by a symbol on the graph. The different features of the symbols illustrated the two different 'happy' variables, and their placement on the graph showed the time of day and the day of the week they occurred.

While Anna's visualization is interesting, Siena's visualization is much more so. In particular, the process and reasoning the second grader went through to draw her final visual illustrated how she grappled with displaying multiple variables in the same visualization.

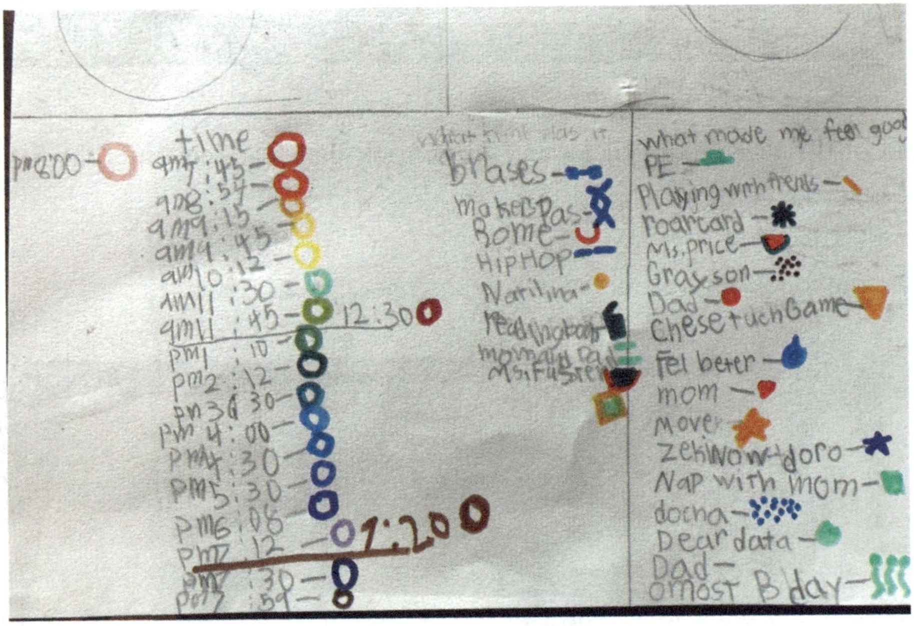

As a first step, Siena began by making her key. Her coding key included every single time she had recorded and every single thing that made her happy. She decided to represent each thing that made her happy with a symbol and then each time with a colored circle. Her idea was to place the colored circle around the symbol to represent when each symbol took place. She decided to make seven circles, each representing a clock for each day of the week, and then place these symbols around the circle.

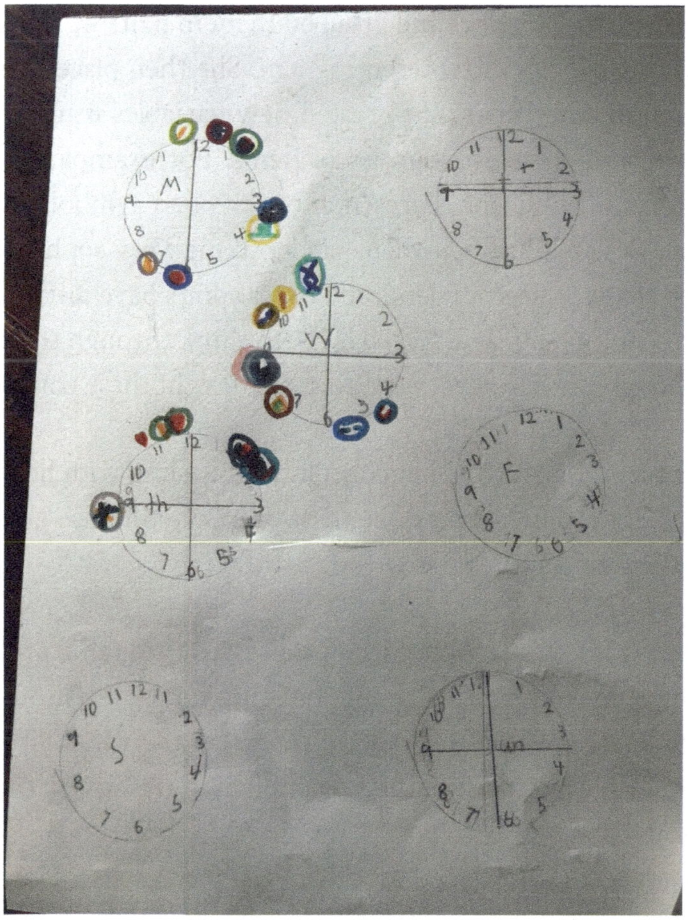

Siena began placing her symbols on the Monday, Wednesday, and Thursday clocks. As she had planned, she drew the symbols and then circled them with the color representing the precise time in her key. This proved to be unsatisfactory to her, and she decided to start again with another display. When asked why her original visualization idea did not work, she stated:

"I can't put *a.m.* and *p.m.* on the clock. And I have too many things, so I don't like it because I don't have room. Plus, the clock already shows the time, so I don't need to circle the things."

In this quote, we hear Siena beginning to think about the multiple dimensions—so much so that she reasons that she cannot represent all of the dimensions she wants in the way she has drawn her graph. In addition, by stating that she has "too many things," she is thinking about how she could possibly group some of her information.

As a final image, she drew 14 clocks and grouped them in pairs—gray-colored clocks representing p.m. and yellow clocks representing a.m. She then placed her symbols around the clock at the appropriate times. She created new variables using the variable "What made Siena happy?" and created categories of items. For example, she grouped friends all into one category; in the prior representation, she had symbols for each friend that had made her happy. This illustrates Siena undertaking very sophisticated multivariate thinking that even many college students and other adults have difficulty understanding. Creating the representation by hand forced Siena to think through these important statistical ideas, even though she has never been formally taught these concepts.

Siena's thought process also showed her constantly grappling with how to best depict the data and capture all of the dimensions of her data precisely. She also grouped the information in ways that she found appropriate.

As demonstrated in this investigation, Siena progressed through a problem-solving process in order to come up with her final graph. In doing so, she questioned the data in ways that helped her understand successful ways to represent the variability in her data across multiple variables. While her questioning was implicit in this investigation, she implicitly answered the following:

- What are the cases in the data set?
- What are the variables in the data set?
- What graphical features are available to use (e.g., color, shape, size, etc.)?
- What graphical features will be used to represent each variable? What patterns are easy to spot in the created graphic? What patterns are not easy to spot?
- How can the graphic be adjusted to emphasize a different feature of the data or show more information?

These data collection and analysis questions are not meant to be an exhaustive list that a teacher may ask students when exploring how to create interesting graphical displays for multivariate data. On the other hand, these questions are meant to illustrate how questioning can be used to understand the data and push the analysis of the data in meaningful ways. Answering such questions allows students and teachers to understand whether their visualizations are successful at uncovering the patterns in the data or not, thus making the graphic more effective or less effective.

This same investigation can be done with students of all levels. For example, here is a beautiful visual created by an undergraduate student. The image represents data collected over the course of five days on when the student checked the time. She noted when, where, why, and how she checked the time.

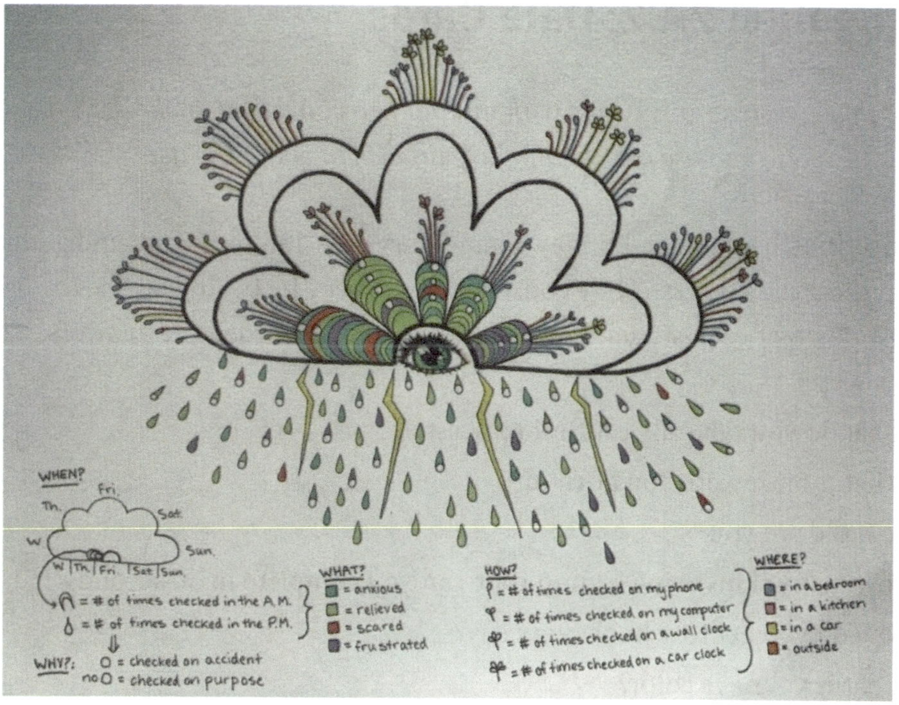

INVESTIGATION SUMMARY:

The main concept developed in the *Dear Data: My week of happiness* investigation is:

Questioning can drive data collection and the process of making a data display. How will this information be best represented? How can we capture the multi-dimensionality of real-life data on a piece of paper? Can the data be grouped? If so, what are appropriate groupings? Answering such questions as one is drawing a data display ensures that the visual will be successful at conveying the information, variability, and patterns in the data.

As we through our daily lives, we take in information about the things and people around us. Sometimes we may formally collect data on these people or objects, recording information about particular characteristics of each thing or person, while other times we have to make sense of data already collected for us. When data are collected for us and presented in a raw form, we have to train ourselves to try to find patterns in the data. This next investigation provides an already collected data set on 42 girls and 34 boys in elementary school. The goal of the investigation is to visualize the data in some way in order to extract patterns from the data.

Investigation 2A.2: Data Cards

Goals of this investigation: Work with multidimensional data, make data visualizations, and use questioning to gain insight about the data.

The entire fourth grade at Plinkey Elementary was interested in understanding the people in their class better. They constructed a survey by posing a series of data collection questions that each student answered, and then recorded the answers. The survey questions were:

- What do you typically eat for breakfast?
- What month were you born in?
- How old are you?
- How many skips using a jump rope can you complete in 30 seconds?
- How do you get to school every day?
- What is your eye color?
- What is your height?
- What is the length of your right foot?

They recorded all of their personal information on a data card. A data card is a card that contains all of the values of the variables included in the data set about a particular case in the data set. In this case, each individual student is a case in the data set, and the variables are the following:

- Breakfast food
- Birth month
- Age
- Number of skips in 30 seconds
- Mode of transportation to school
- Eye color
- Height
- Length of right foot
- Grade

Each of these variables matches one of the survey questions.

The students agreed on a format for their data cards. The format illustrates where each of the students will record the information about themselves on the card. Here is a picture of the format:

Variables:

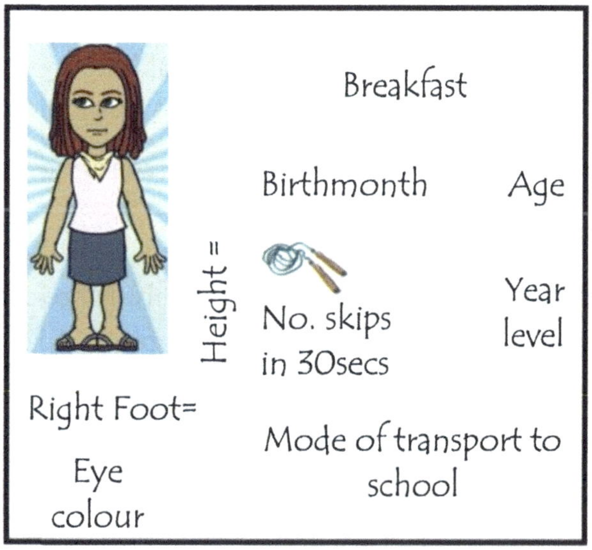

Using this format, each child created their own data card by hand or on a computer using the template. Here is an example of Siena's data card, written by hand:

There are 76 fourth grade students at Plinkey Elementary. Of these, 42 are girls and 34 are boys. Each student created their own data card. (The full set of data cards can be printed from Datacards.pdf.)

A series of investigative questions was then posed by the students:

- What month are students in fourth grade typically born?
- What is a typical amount of skips we can expect fourth grade students to do in 30 seconds?
- What do students typically have for breakfast?
- How do students typically get to school?
- What color eyes do students in fourth grade typically have?

Note that all of these questions involve only one variable. In other words, students have to analyze the data according to one variable only. More complex questions that students could pose include multiple variables. For example:

- Do girls or boys typically eat different things for breakfast?
- Does what you eat for breakfast influence how you get to school?
- Are people who eat breakfast able to skip more than those who don't?

Using the data cards, students can arrange them in graphical displays that will help answer their posed questions. For example, Janelle examined a subset of 18 students and posed the following question:

In what month were the students in our class typically born?

By arranging the data cards into groups according to their birthdays, Janelle created the following data display to help answer her question:

In this display, Janelle represented the people in the group of data cards that she was using who were born in each month. This provided a summary of the data instead of showing individual cases or people who were in the class. This could be considered a pro or con for this display. On one hand, having the summary makes it easy to compare frequencies across months; on the other hand, the individual representation of each case is lost in this graph. Janelle organized the data around a circle, representing the cyclical nature of the months in a year. From this display, Janelle answered the investigative question as:

"The months of October, November, and January appear to be the most common birthday months for kids in our class. Overall, fall and winter birthdays are more common than summer and spring birthdays."

Chen, another student, chose a different question to investigate:

What are the typical eye colors of students in our class?

Chen visualized the class data in the following manner:

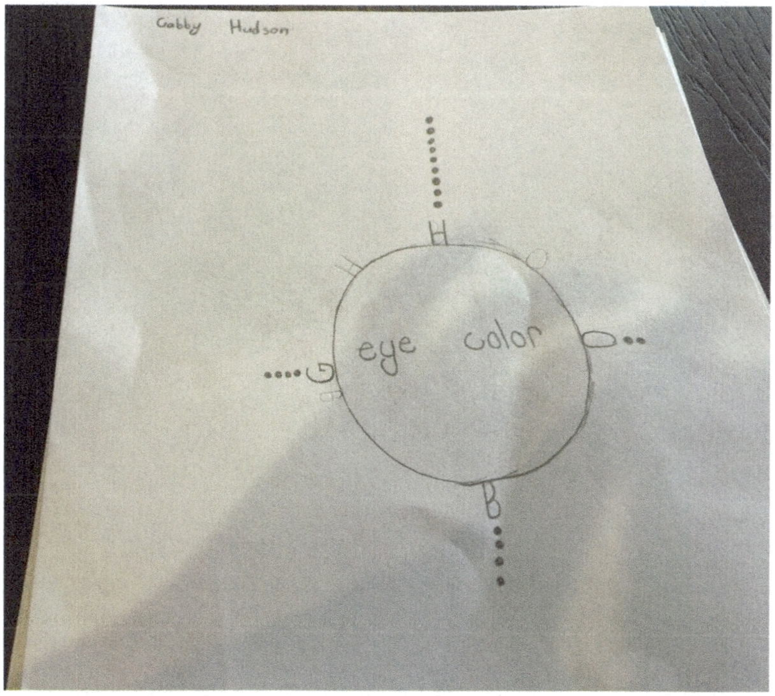

Although this graph is creative, it is difficult to determine which colors were more frequent than others. By having the data represented as dots around a circle, it is difficult to make a direct comparison between categories of eye color if the number of students with each category of eye color is close. In this case, the H category, standing for "hazel eyes," was the most common; however, without counting the dots, it is hard to see whether green or blue were the same or whether one was larger than the other. Chen's graph represented each individual on the graph and did not summarize the data in any way. While Chen also placed his categories around a circle, in this case, the circular representation is distracting because, unlike months in a year, there is nothing circular about eye color. Use of color to represent the variable *eye color* might have been a more effective way to convey the information. Having the eye colors be color-coded would make it much easier and quicker to understand the graph. In addition, the O category is not well defined on the graphic, leaving one to wonder what type of eye color this category might represent. (The O category stood for "other," but it is unclear what such other colors might be).

Another interesting graph was developed by Shawn, who visualized the type of breakfast students ate. Each leaf on the tree represented a person in the class, and the leaves were color-coded to illustrate the type of breakfast the students ate. A key was provided to show what the different colors represented. At first glance, we can see that cereal and pancakes were the most popular breakfast choices in the class. Similar to Chen's graph, however, it is difficult to compare those two categories without counting the leaves.

Students in the class could create graphs that involved more than one variable. For example, Shawn's graph could be split into two trees, one for boys and one for girls. This display could help shed light on the answer to the following question:

Do girls or boys typically eat different things for breakfast?

Another interesting component of using the data cards to collect information and then summarize the data in a graphical display is that students need to determine how to put their data into a proper format. For example, when students were asked what they ate for breakfast, some might respond with "Cheerios" and others might respond with "cereal." When making a visualization, a student or teacher must choose whether to cluster all the cereals into one group—such decisions could dramatically affect the final graphics.

The initial task of making a data card highlights the fact that each individual in a class is a "case" in the data set. Because the students are required to collect multiple pieces of information on themselves, they begin to think about multiple variables at a time. Students can then be pushed to think about how to represent multiple variables in one graphical display. By first looking at single variables and providing visuals to help make sense of the data, students can then graduate toward making visuals to represent relationships between variables. This is the beginning of multivariate thinking at an early age through data visualization. Drawing visualizations by hand also forces students to think about every aspect of their cases and representations of those cases.

INVESTIGATION SUMMARY:

The main concepts developed in the data cards investigation are:

1. Many of the data representations we see in our daily lives are not those that we study in school.
2. Questions can be posed that relate variables to one another.
3. Creativity is often essential in creating a good visualization of nontraditional data.
4. Multiple variables can be represented in the same visualization.

Follow-Up Questions

1. Create your own *Dear Data* investigation.
2. Create your own survey data cards that would be appropriate for the grade levels you teach. Pose an investigative question and use the data collected with the data cards to help create a visual to aid in answering the question.

Students can be encouraged to explore creative graphical visualizations and then pushed to think about how helpful the graph is for answering the posed question, as well as about the overall pros and cons of each display. This exercise of allowing students to push the boundaries of representation can prove to be difficult. As mentioned, many of the data representations we see in our daily lives are not those that we study in school (e.g., bar plots, histograms, etc.). Instead, often we see maps, word clouds, and many other interesting and nontraditional data representations. The context dictates the effectiveness of the graph.

References

Börner, K., Bueckle, A., and M. Ginda. 2019. Data visualization literacy: Definitions, conceptual frameworks, exercises, and assessments. *Proceedings of the National Academy of Sciences of the United States of America* 116(6): 1857–64.

Díaz, A. 2020. Discover the four V's of Big Data. Open Sistemas, June 16. https://opensistemas.com/en/the-four-vs-of-big-data.

King, A., Chew, N., Jay, A., MacLean, A., and A. Bargagliotti. 2021. A guide to modern data visualization. *Math Horizons* 28(1): 24–7.

UNIT 2B:
Toward Data Science

Data do not necessarily come in a well-formatted spreadsheet or table. Instead, data might be presented in nontraditional forms such as text, pictures, sound bites, etc. and might be organized by cases (equivalent to the idea of a data card). Such nontraditional data might require unconventional graphical displays which provide visual displays on multiple variables. Nontraditional data might also require alternative types of analyses for summarizing the distributions.

Unit 2B includes two investigations. In the first investigation, students explore the idea of using pictures as data to answer a posed statistical investigative question. Students interrogate the picture data to define multiple variables, record the variables into a table or spreadsheet form, and utilize interactive software to help visualize the data and answer the investigative question. In the second investigation, students examine a map pictured in What's Going On in This Graph? (www.nytimes.com/column/whats-going-on-in-this-graph), create their own data set based on the interactive map, create their own graphical displays to answer specific investigative questions, and then draw appropriate conclusions.

Investigation 2B.1: Pictures as Data About Us[3]

During the COVID-19 pandemic, many students around the world experienced school from home in the spring and fall of 2020 and into 2021. A sixth grade class wondered about students' workspaces at home and asked the following statistical investigative question:

What do typical home workspaces look like for students in our class?

To answer this statistical investigative question, students in the class decided to collect primary data (data collected by the researcher) by having everyone in their class take

3 This investigation is written up in Bargagliotti, A., Arnold, P., and C. Franklin. 2021. GAISE II: Bringing data into classrooms. *Mathematics Teacher: Learning and Teaching PK–12* 114(6): 424–35.

pictures of their home workspaces and add them to a shared class folder. In addition, every student was asked to answer the following survey question:

How does your workspace make you feel?

Students were asked to answer the survey question using a maximum of 180 characters. Along with their picture, students submitted the comment about their picture in a text document.

The pictures from one small class of 20 students are included in the WorkStationPictures folder.

Once the data were collected, students formulated further data collection questions to define and collect data on multiple variables from the photos. For example, the sixth grade students generated the following data collection questions about the pictures, which resulted in defining 11 variables:

1. What type of surface is the workspace?
2. How many screens are there at the workspace?
3. Is there a lamp at the workspace?
4. Are there colored pencils and pens at the workspace?
5. Are there books at the workspace?
6. Are there binders at the workspace?
7. What types of computers/devices are there at the workspace?
8. How many computers/devices are there at the workspace?

9. What is one word that describes the workspace?
10. What is the dominant color of the workspace?
11. How many objects are plugged in at the workspace?

A data card can be made from each printed picture to help students understand that the unit of observation (a case) is the workspace. Each student can place the answers to the 11 data collection questions on the back of their picture, as illustrated in the following photos for one of the sixth grade students.

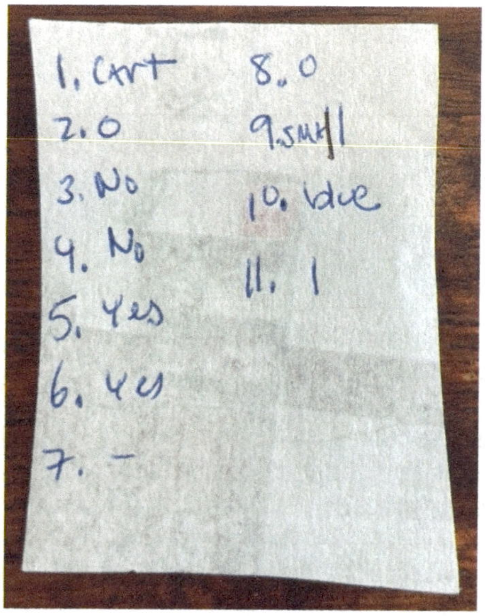

The class data can then be put into a spreadsheet using some type of interactive medium, such as Google Sheets (see the following image for the sixth class). Students at this point should recognize that one row in the spreadsheet represents one student workspace, namely, one case. Multiple variables are collected for each student workspace.

	A	B	C	D	E	F	G	H	I	J	K	L
1	What type of surface is the workspace?	How many screens are there?	Is there a lamp?	Are there colored pencils and pens?	Are there books?	Are there binders?	What types of computers/devices are there?	How many computers/devices are there?	One word to Descibe it	Dominant Color	Number of items plugged in	Makes you feel
2	Cart	0	No	No	Yes	Yes		0	small	blue	1	So organized!
3	Table	1	No	Yes	No	Yes	Chromebook	1	tablecloth	blue	1	I have too much stuff
4	Table	1	No	Yes	Yes	No	Chromebook	1	small	yellow	1	I don't like learning at home
5	Table	1	No	Yes	No	Yes	Chromebook	1	fun	white	1	My desk looks so inviting!
6	Desk Cart	1	Yes	Yes	No	Yes	Macbook	1	pretty	white	3	My desk is so pretty!
7	Desk	1	No	Yes	Yes	No	Chromebook	1	cluttered	white	1	I work well here.
8	Desk	1	Yes	Yes	No	No	iPad	1	simple	white	3	I really don't like working at home. It makes me sad.
9	Desk	1	No	No	No	No	PC	1	neat	white	1	I am very organized with my stuff
10	Desk	2	No	Yes	No	Yes	Chromebook and PC	2	organized	white	4	I feel happy when looking at what I have done.
11	Desk	2	No	No	Yes	No	Chromebook and PC	2	small	pink	2	I am accomplished.
12	Desk Cart	2	No	No	No	No	PC and iMac	2	small	blue	3	I like to work on my own at home.
13	Desk	2	Yes	Yes	No	No	Chromebook and iMac	1	clean	white	3	My desk is really nice.
14	Desk	1	No	Yes	Yes	No	Chromebook	1	pretty	pink	1	My desk shows who I am!
15	Cart	0	No	No	No	No		0	small	green	1	I just put my stuff on the cart!
16	Desk	1	Yes	Yes	No	No	Chromebook	1	cluttered	white	1	So much stuff everywhere!
17	Table	1	No	No	No	No	Chromebook	1	simple	green	1	I feel overwhelmed.
18	Desk	1	No	No	No	No	Chromebook	1	simple	yellow	1	I feel sad without my school friends.
19	Desk	1	No	No	Yes	No	Chromebook	1	neat	blue	1	I love to work at home at my awesome desk!
20	Desk	1	No	Yes	No	No	Chromebook	1	neat	blue	1	My desk makes me really happy because I can organize
21	Table	1	No	Yes	Yes	No	Chromebook	1	cluttered	white	1	I just try and g4t my work done quickly!

To investigate the characteristics of the typical student workspace, students must consider all of the variables at once. Multivariate thinking is intuitive and natural for students; thus, students should be encouraged to develop analysis questions that will help guide their exploration. In the following table, the data collection questions and the defined variables are listed with potential analysis questions students could develop as a group. The questions guide students to discuss each variable and suggest strategies for analyzing the data.

Data Collection Question	Variable	Analysis Questions
1. What type of surface is the workspace?	Type of surface	What is the most common surface? What is the least common surface? How many different types of surfaces are there?
2. How many screens are there at the workspace?	Number of screens	How many students have at least one screen? What proportion of students have at least one screen? What is the most common number of screens? Are there any unusual responses?
3. Is there a lamp at the workspace?	Has a lamp or not	Is it more common to have a lamp or to not have a lamp? What proportion of students have a lamp?
4. Are there colored pencils and pens at the workspace?	Has colored pencils and pens or not	How many students have colored pencils and pens? What proportion of students have no colored pencils and pens?
5. Are there books at the workspace?	Has books or not	Is it more common to have books than to not have books? Are there any unusual responses?
6. Are there binders at the workspace?	Has binders or not	Is it more common to have binders or not?
7. What types of computers/devices are there at the workspace?	Types of computers/devices	How many different types of computers/devices are there? Which type of computer/device is most common? What proportion of students have the most common computers/devices? Which type of computer/device is least common? Is there a lot of variation in the types of computers/devices that students have?
8. How many computers/devices are there at the workspace?	Number of computers/devices	What is the maximum number of computers/devices? What is the minimum number of computers/devices? What is the median number of computers/devices? What is the mean number of computers/devices? How much variation is there from the mean?
9. What is one word that describes the workspace?	Description of the workspace (one word)	What word is most used to describe the workspaces? Are there any unusual words used? How many different descriptions are there?

Data Collection Question	Variable	Analysis Questions
10. What is the dominant color of the workspace?	Dominant color of workspace	What is the most common dominant color of the workspaces? What is the least common dominant color of the workspaces? Are there any unusual dominant colors of workspaces?
11. How many objects are plugged in at the workspace?	Number of objects plugged in	What is the median number of items that are plugged in? What is the mean number of items that are plugged in? How much variation is there from the mean?

To address the analysis questions listed in the table, graphical displays can be developed that help students identify patterns and connections in these data. Such displays can be constructed by hand, organizing the pictures into bar graphs or dotplots, or using technology. For example, to help with analyzing the data collected for type of surface, students produced the following graphical display using their photos which shows that a large number of students worked at a desk:

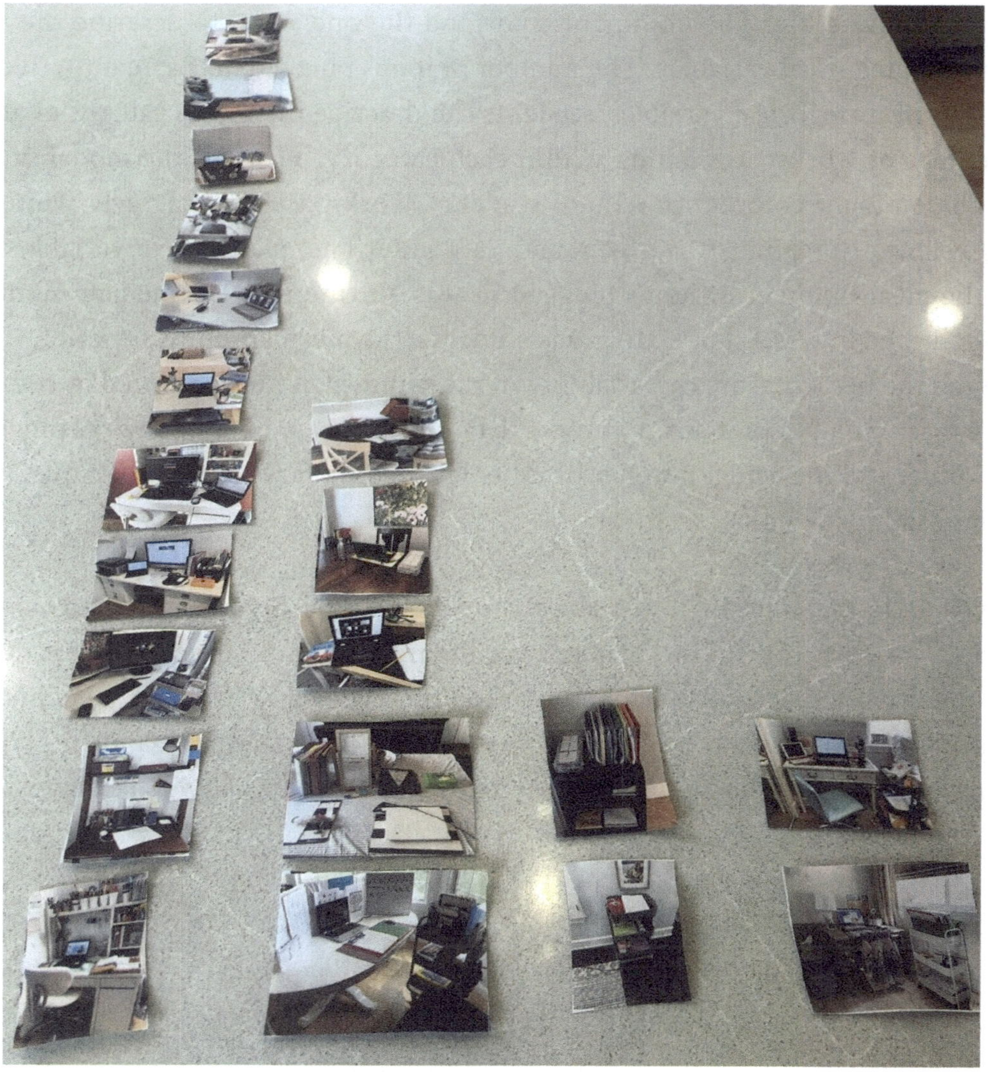

The students used software to create graphical displays. From the following bar graphs, students could visualize that the majority of students in their class are working at a desk; five students are working at a table, and only two students are working at a cart. The modal category is a desk. Graphical displays can be made using free online data-analysis platforms such as CODAP (https://codap.concord.org):

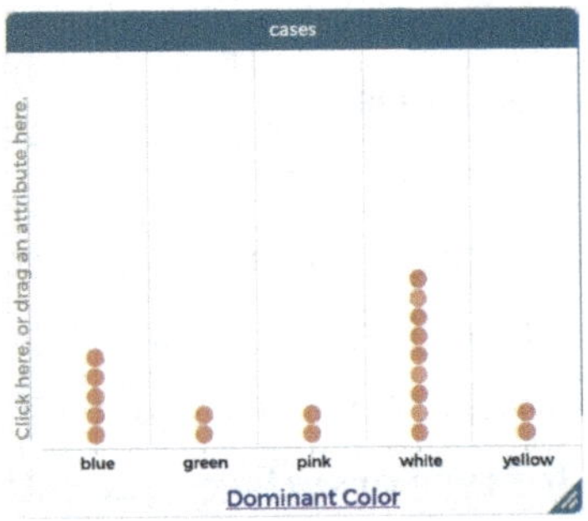

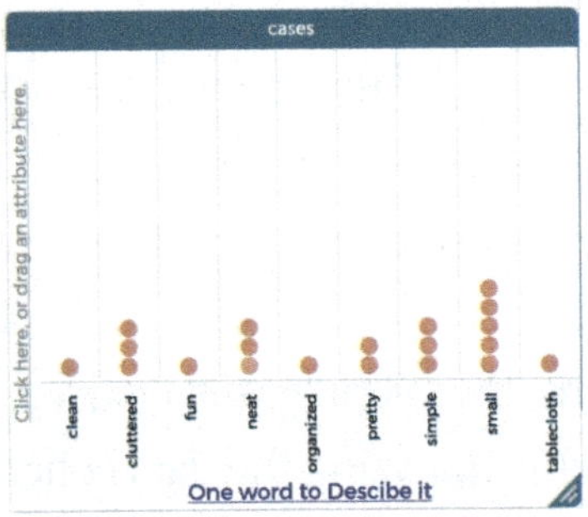

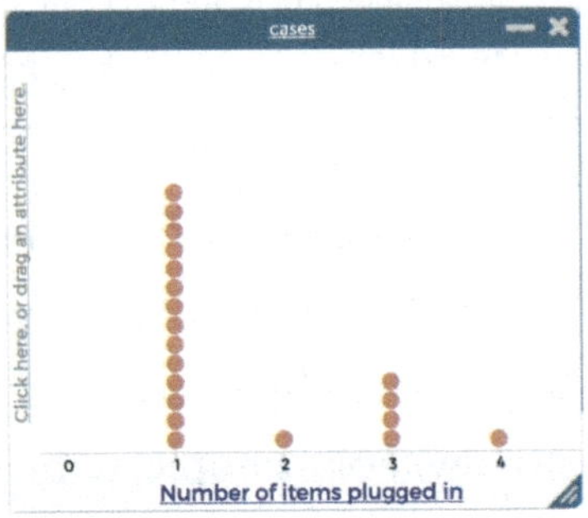

These CODAP visuals display the distributions of two categorical variables and one quantitative variable. The dominant color and the one word to describe the workspace are categorical variables. The number of items plugged in is a quantitative variable. For the categorical variables, students can describe the modal categories and the percentage of workspaces falling within each category. White is the modal category for the dominant color, with 45% of students' workspace being largely white; 10% were yellow, 10% pink, 10% green, and 25% blue. The quantitative variable shows that the mean number of items plugged in was 1.6 and that the median number of items plugged in was 1. For sixth grade students, the mean absolute deviation (MAD) provides the best measure of variability in the number of items plugged in compared to the mean number of items. For these data, the mean number of devices plugged in is 1.6 units and the actual number of units plugged in vary from 1.6, on average, by .84 units (almost 1 unit).

As well as looking at single variables, students should be encouraged to consider multiple variables together to help answer the investigative question "What do typical home workspaces look like for students in our class?" For example, students could explore the association between the types of workspace surfaces and the types of computers/devices. This figure for the sixth grade students shows the association as a two-way table.

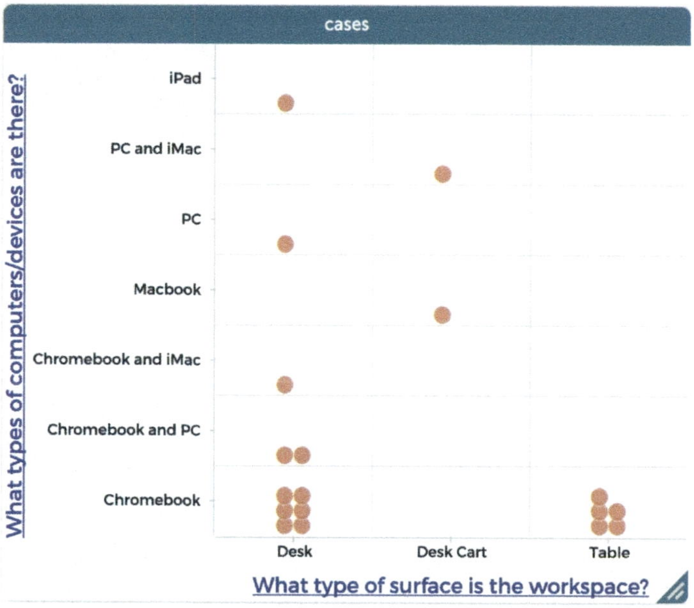

Some example analysis questions students could ask of this graphical display are:

1. What is the most common combination?
2. Are there some combinations that do not exist?
3. Does all of one category from one variable match up with all of one category from the other variable?
4. What variability is there within a category?

The two-way table shows that Chromebooks are most popular in both the desk and table surfaces. In fact, nine out of the 20 workspaces have Chromebooks at their desks, and five out of the 20 workspaces have Chromebooks at a table. At the tables, 100% of devices are Chromebooks. Three of the people who have Chromebooks also have either a Mac or a PC at their desk.

Students could push their investigative questions further by looking at associations for more than two variables. Students can grapple with how to explore associations among more variables by representing multiple variables on the same graphic, first by hand. For example, the following display from the sixth grade class shows the association among dominant colors of a workspace, the number of computers/devices, the number of items plugged in, and the surface of a workspace.

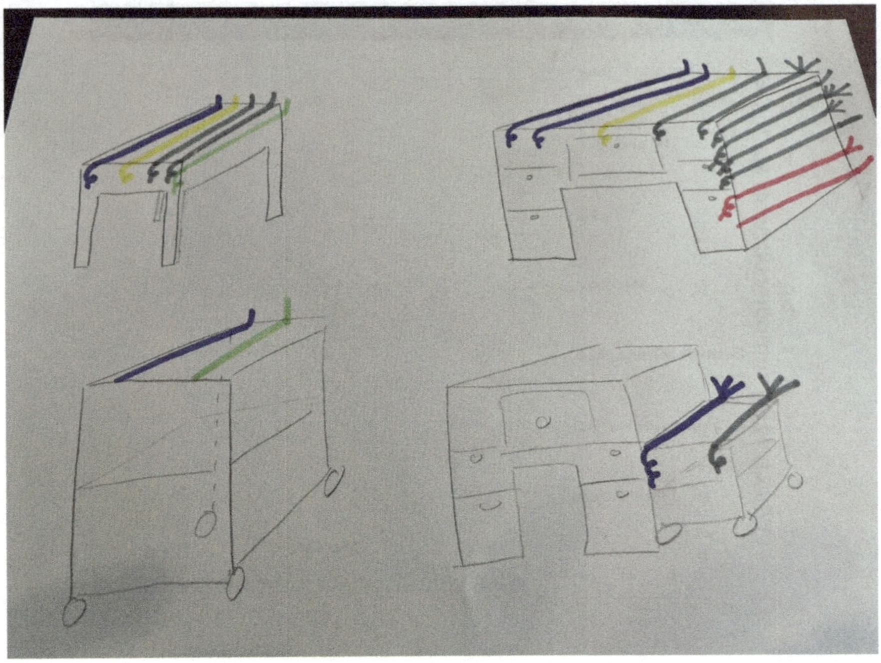

The workspace surfaces are sketches of a table, a desk, a cart, and a desk and cart. On each surface, a line represents the dominant color for a workspace. There are five colors: blue, green, yellow, pink, and white (white is shown as grey in drawing). The number of bars stemming off from the colored line (at the top) represents the number of items that are plugged in, and the number of squiggles (at the bottom) represents the number of computers/devices at the workspace. With this graphic, we can see that desks are the modal surface category, that the table surfaces have only one item plugged in on each, and that the cart surfaces do not have any computers/devices. This could be because a table might have an overhead light; thus, a student working on such a surface might need only their device plugged in, and a student at a cart may keep their devices in other places. To compare, consider the desk images: These images have several cases in which multiple things are plugged in. This is also true of the desk and cart. Also the dominant color of the desks is white, and pink appears only at the desk surfaces. The distribution of colors has the most variability for the desks. This could be because a desk is a personal space in which students may have the freedom to introduce whatever colors they prefer. Many other observations can be drawn from the graphic. Using hand drawings is important for students because it encourages them to think about different ways that multiple variables can be included in a multidimensional graphic.

Another example of a graphic that represents three variables is given in the next drawing. In this graphic, the sixth grade student represented the number of screens, the surface type, and the dominant color. In the bottom left cell—cart/zero screens—the student stacked the colors blue and green on top of each other. In the other cells, the student distinguished the color into different bars. Students should recognize how stacking the colors in different bars makes for a better graphic because the distribution of colors can then be more easily compared within a cell. The relationship among these variables can also be explored using free online software, such as Tableau Public (https://public.tableau.com/en-us/s/).

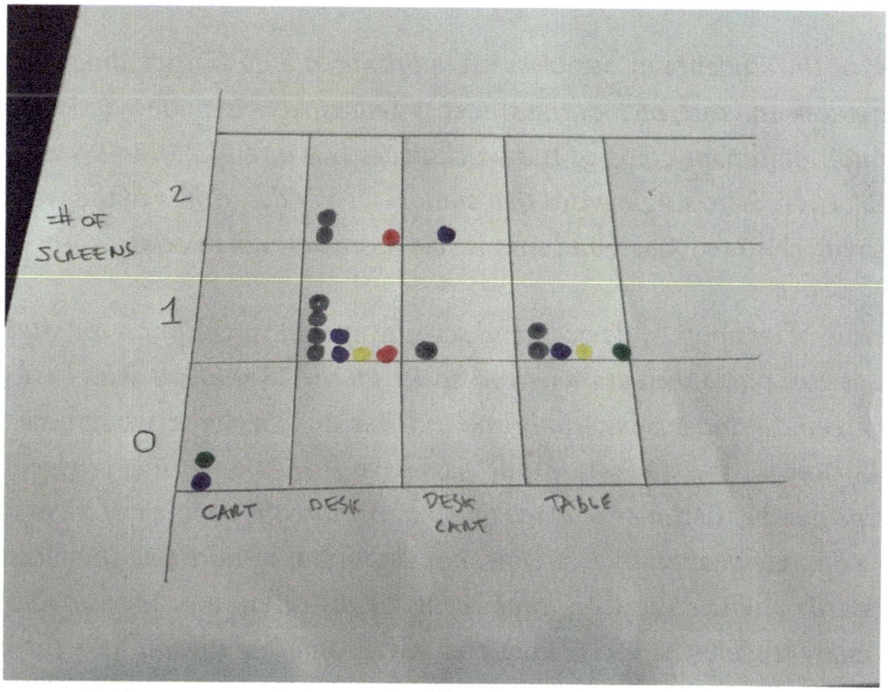

Students synthesize the analyses into a response for the statistical investigative question "What do typical home workspaces look like for students in our class?" The teacher could extend students' thinking further by asking the following statistical investigative question: "What do typical home workspaces look like for students in our grade?" Can we use the data from our class to answer this investigative question? Students should note that in this case, the class would be a sample. If a different class's pictures were selected, the images would vary. Students should recognize the limitations of any conclusions that they draw from these data to answer the investigative question for *students in their grade.* Students should be encouraged to contemplate what spaces might look like if they had a different sample of pictures from another class and how the variables they defined might vary depending on the pictures in the sample.

The statistical investigative question, data collection questions, and analysis questions outlined in the investigation are by no means an exhaustive list. Students should be encouraged to develop many more that utilize all of the other variables (e.g., the 180-character description). To answer the statistical investigative question, students are encouraged to make visuals, because these can help provide evidence and support their analyses, as well as use analysis questions to help make sense of their visuals, because this helps students think about and statistically reason with multiple variables at a time. In turn, the answers to the analysis questions asked will help provide an overall picture to answer the investigative question about describing the typical workspace. An example answer to the investigative question for the sixth grade class,"What do typical home workspaces look like for students in our class?," is:

> *Most of the students in our class are working at a desk. Including those who have a desk and cart, 65% of the students' workspaces include a desk. The most common dominant color of the workspaces is white (45%), followed by blue (25%). There is no single word that stands out as a descriptor of the workspace, but* small, cluttered, neat, *and* simple *had more votes than other words.*
>
> *In terms of computers, devices, and screens, it is most common for students to have at least one screen at their workspace. Of the 18 students who have at least one screen, 14 students have only one and four students have two screens. All the workspaces for the students in our class have at least one thing plugged in; the average number of things plugged in is 1.6. Because the number of items plugged in is a discrete quantitative variable, it is important to note that the mean is not necessarily a whole number. Looking at the distribution of the variable shows that many students have one item plugged in, and that several have three items plugged in. Exactly five students have three or four items plugged in. Altogether, 14 students have Chromebooks, and two students have no computer or device in their workspace.*

This serves merely as an example of a student write-up that includes only the variables mentioned in the previous graphical displays. However, students completing this investigation may choose a different combination of variables to answer the question (such as a book, colored pencils and pens, binders, type of workspace, and types of computers/devices), and in fact should be encouraged to utilize all variables in the data set.

INVESTIGATION SUMMARY:

The main concepts developed in the pictures as data about us investigation are:

1. Multiple variables can be defined based on one picture. Data-generating questions can be asked of pictures, and multiple variables can be recorded for a single case.
2. Visual representations can use color and shape to represent multiple variables in the same graphic.
3. Questioning can guide not only the data generation process, but the analysis process as well.

Follow-Up Question

1. Do the same investigation using pictures of dinners, lunches, or the front door of a student's home.

Technology and computational thinking is an important part of moving students toward modern statistics and data science. In this investigation, students have used technology in many ways: They have taken a picture, uploaded this picture into a data repository (a shared folder), and used software to analyze the data. In addition, they have defined variables, created a spreadsheet, formed new variables from existing ones—all important concepts that require computational thinking.

Investigation 2B.2: Climate Change in Our Community[4]

Climate change is one of the most important challenges of our time. Around the world, extreme weather presents unprecedented structural and economic challenges for humans. Students in a sixth grade class took part in a comprehensive investigation of climate change, reading opinions about climate change and trying to understand climate change in their own environment, writing an essay, debating, and looking at data. This investigation presents the data portion of a broader unit on climate change for students to participate in. Students in this sixth grade class specifically wondered how the personal environments in their community were being affected by climate change. The students posed the following investigative question:

What are typical climate challenges that affect our community?

To investigate this question, the sixth grade students used multiple data sources, including an interactive map found in *The New York Times*, data they collected based on the map, and a photo-voice project (Herrick and Gralnik, in progress) that captured perceptions of climate change through their eyes.

Thinking Like a Data Scientist:
A Cross-Disciplinary Investigation on Climate Change

To begin with, students participated in a data talk as designed in the What's Going On in This Graph? collaboration between the ASA and the *NYT*. Students were presented with

4 This investigation was developed by Anna Gralnik, a fifth grade teacher at Aurelia Pennekamp Elementary School, in California. Bargagliotti, A., and A. Gralnik. Forthcoming 2021. Thinking like a data scientist: A cross-disciplinary investigation on climate change. *The Statistics Teacher.*

the following map and asked to discuss what they observed on the map. To guide their observations, the students were provided with a set of questions (those used in What's Going On in This Graph?) to address when discussing issues related to climate change:

- What do you see on this graph?
- What do you notice?
- What do you wonder?
- What impact does this have on you and your community?

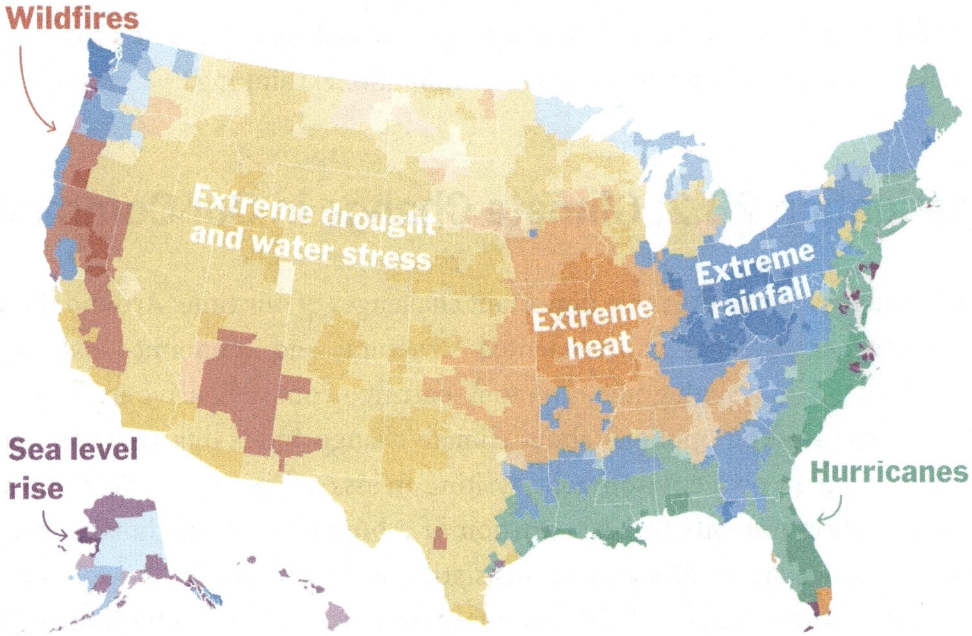

Students can discuss these questions in breakout groups. Teachers should expect that when students are left to discuss, they will comment on all parts of the United States. Teachers can direct students to think about their notion of community on the macro level of the overall United States, and then hone in on the smaller, micro levels of their state and even their county. Teachers can encourage students to focus specifically on their own geographical area, given that the investigative question is specifically about their own community. Here are some examples of answers from these students:

The colors on the chart make sense to what they are representing. Yellow is like the Earth without water or like the desert; blue is like water, for rain; and red is like fire.	Our community is mostly really dry, which makes us prone to wildfires. Most of it is basically a desert and really hot. We have warm waters and lots of drought. We also have some places that have heavy rainfall.	Since the sea levels are rising, people who own property by the beach might get flooded, so they would be forced to move in-state

Six distinct colors are represented on the graphic. Each color represents a type of environmental risk. The risk variable is categorical, with six categories. The categories are:

wildfires, sea level rise, extreme drought and water stress, extreme heat, extreme rainfall, and hurricanes. Students should notice that different areas of the United States are subject to different environmental risks. Large portions of the United States are subject to water stress. Extreme heat is focused mostly in the mid-South and lower rust belt. Extreme rainfall is in the Northeast, and hurricanes hit the Atlantic Coast and southeastern states. The Pacific West Coast has some extreme rainfall in the Northwest, as well as wildfires.

After observing the static map, students were asked to interact with the same map. Students were provided with the link and asked to move their cursor to their state: www.nytimes.com/2020/10/15/learning/whats-going-on-in-this-graph-climate-threats.html.

To guide their data investigation with the interactive map, the teacher posed the following analysis questions to help students begin exploring; they can be used as examples of questions to use with other students:

1. What does the color-coding represent?

Students ideally should have noticed this already in the static map, in which the environmental risk is a categorical variable with six categories. An example student answer might be:

> *Each section of color had a description that usually started with extreme, which means that climate change is serious and dangerous.*

Teachers should encourage a student to articulate that there are six colors, each representing a different type of risk.

2. What does the darkness/lightness of the color represent?

The gradation of the color represents how severe the risk is. The darker the color, the greater the risk. For example, although both Missouri and Tennessee can experience extreme heat as an environmental risk, Missouri has a higher risk.

3. Where did the data come from? What data were collected to make this map?

To understand the source, students should be guided to the original article presenting the map in the *NYT*: www.nytimes.com/interactive/2020/09/18/opinion/wildfire-hurricane-climate.html.

The original article cites that the data came from Four Twenty-Seven (http://427mt.com/), a company that focuses on assessing climate risk for financial markets. The data for the map were taken from one of the company's reports: http://427mt.com/wp-content/uploads/2018/05/427-Muni-Risk-Paper-May-2018-1.pdf.

A student in this sixth grade class wondered the following:

> *How do people collect and track the weather issues in different places related to the risks?*

Data have a source, and the source needs to be checked by students to make sure it's reputable. Reputable sources have citations. They have a description of their data collection process and describe any limitations of the data. Data collection and study design are advanced and extremely important parts of statistics. It is essential that students at an early age develop the practice of checking sources as a necessary process when drawing conclusions or making any assertions. This process should not be undervalued or skirted. It is a crucial part of the statistical investigative process that falls under the Collect Data component.

4. What are you noticing now?

Now that students can interact with the map, they can notice that as their cursor hovers over the map, more detailed information is provided, including data at the county level. Students can also see that as they hover over a county, each county has been assigned a category for each of the risk types. This information allows students to make more specific observations about the data as it relates to individual counties. For example, here is a student observation made after the student engaged in an "inquiry center." An inquiry center is a center where students examine different readings, photographs, writing assignments, and discussion topics. The teacher in this sixth grade class set up four such centers. A student from one of the centers made the following observations:

> *I noticed that in our inquiry centers, we saw that in the photography inquiry center, there was a wildfire and Los Angeles had medium wildfire risk. In the writing inquiry center, we saw a drought, which had a very high risk where we live. This can show that climate change can affect more than one place or person.*

This student connected the information they found in two inquiry centers to the data they saw on the map. The map shows that each area can have multiple risks, and sometimes the risks may be connected (e.g., wildfires and drought).

5. Explain what features of the interactive map help you to notice.
Here are additional observations made by students:

> *The darker the color is, the more extreme the weather, but the lighter the color, the less extreme the weather will be. Also we noticed where the most crises are affecting other places, not just one location.*

6. What conclusion can you draw based on your discoveries?
Here in an example of a student's conclusion:

> *Water stress or wildfires + rain make mudslides/landslides. So in CA, we have water stress, wildfires, and rain. When wildfires burn the plants or when water stress makes them dry, their roots can't hold the soil together. Then when it rains, the water washes through the soil, making it watery and turning it to mud.*

7. Continue exploring and noting your discoveries.
Students can begin to turn their observations into more complete and coherent thoughts. For example, one group of students decided to try to connect some of the different risks. By hovering over a county on the map, students could see that for each of the risks, the county is given a rating of very high, high, medium, low, or no risk. These ratings are what determine the darkness of the color. After looking at several counties, one group developed explicit relationships among the different risks. They said:

- *Water stress + rain = landslide*
- *Extreme heat + water stress = wildfire*
- *Rising sea levels + rainfall = floods*

Students should be encouraged to show evidence for these statements to support their conclusions.

8. What are you noticing about your own community?
Student answers will vary to this question. For example, a student shared:

> *Weather will have an impact because if we don't solve this problem, it will probably [have a bigger impact on] our community.*

Because the investigative question focuses on the students' local community, they should be encouraged to focus on their area of the map. In addition, students in the sixth grade class were able to connect the data displayed in the map to the prior inquiry centers. One student referred back to the inquiry center where they compared photos of the same place a few years apart and shared:.

www.santa-ana.org/pw/water-and-sewer/water-services/conservation/californias-drought

I noticed that the right side of the picture has very cool colors and the left side looks like it is very hot. So it shows that it is getting hotter. Also, I saw that it was very green and a nice big lake but in 2014, the lake is about three times smaller, and the surrounding greenery is no longer there or very dry and brown.

9. What are you noticing about other states in the United States?

Student answers will vary for this question. Here are three examples of what three different students said:

- *On the map, Hawaii was all purple, which means the sea levels rise. Maybe in a couple of decades, the island might be under water.*
- *Hawaii seems to be covered by rising sea levels, and since it is not too far from us, it can affect us by pushing the waves over, putting us in a position of rising sea levels too.*
- *We compared Arizona and California. We discussed whether Arizona or California had more wildfires. We decided that Arizona has more wildfires, but California has more severe cases of wildfires because California has more vegetation.*

Once students finished interacting with the map, they were given time to draw conclusions and summarize their findings in the greater context of the investigation. As part of their class, as mentioned previously, students had read articles discussing climate change, read different opinion pieces about climate change. They also had to write an essay about their thoughts on climate issues relating all of the resources they studied. Unprompted, students created debates between climate-change activists, such as Greta Thunberg, and climate-change deniers, such as Naomi Seibt. They consolidated the conclusions they were making when talking through the answers to questions 1–9 in groups, thus focusing on communication with data (an important principle of data science and statistics).

It depends where you live. The specific place will determine what type of weather issues you'll get.

Next, students were then asked to dive deeper into climate change in their own community, not just their region, and create a data set that shows the different climate threats and to what degree the threat is around them. Each group of students was assigned 10 counties in California and asked to use the map to collect data on those counties. Here is an example spreadsheet created by one group of students:

	A	B	C	D	E	F	G
1		**Extreme Rainfall**	**Water Stress**	**Heat Stress**	**Wildfire Risk**	**Sea Level**	**Hurricanes**
2	Sierra	Medium risk	High risk	Low risk	Very high risk	No risk	No risk
3	Siskiyou	Medium risk	High risk	Low risk	High risk	No risk	No risk
4	Sedona	Medium risk	Medium risk	Medium risk	Medium risk	Low risk	No risk
5	Sonoma	Medium risk	Medium risk	Medium risk	High risk	Low risk	No risk
6	Stanislaus	Medium risk	High risk	Medium risk	Medium risk	No risk	No risk
7	Sutter	Medium risk	Medium risk	High risk	Medium risk	No risk	No risk
8	Tehama	High risk	Medium risk	Low risk	High risk	No risk	No risk
9	Trinity	High risk	Medium risk	Low risk	High risk	No risk	No risk
10	Tulare	Very high risk	Medium risk	High risk	Low risk	Low risk	No risk
11	Tuolumne	Medium risk	High risk	Low risk	High risk	No risk	No risk
12	Ventura	Low risk	Very high risk	Low risk	Medium risk	Medium risk	No risk
13	Yolo	High risk	Medium risk	Medium risk	Medium risk	No risk	No risk
14	Yuba	High risk	High risk	Medium risk	Medium risk	No risk	No risk

For each county, students recorded the severity of each of the risks. On their own, students then tallied the risks into a table and made additional graphical displays, such as the following bar and pie graphs, to help illustrate the risks in their local community. From the data collected on these counties, the students made a summary table of the data:

	Extreme Rainfall	**Water Stress**	**Heat Stress**	**Wildfire Risk**	**Sea Level**	**Hurricanes**
Very High	1	1	0	1	0	0
High	4	5	2	5	0	0
Medium	7	7	5	6	1	0
Low	1	0	6	1	3	0
No Risk	0	0	0	0	9	13

Based on the summary table, they were able to create graphical displays to help them understand the severity of the risks in their counties.

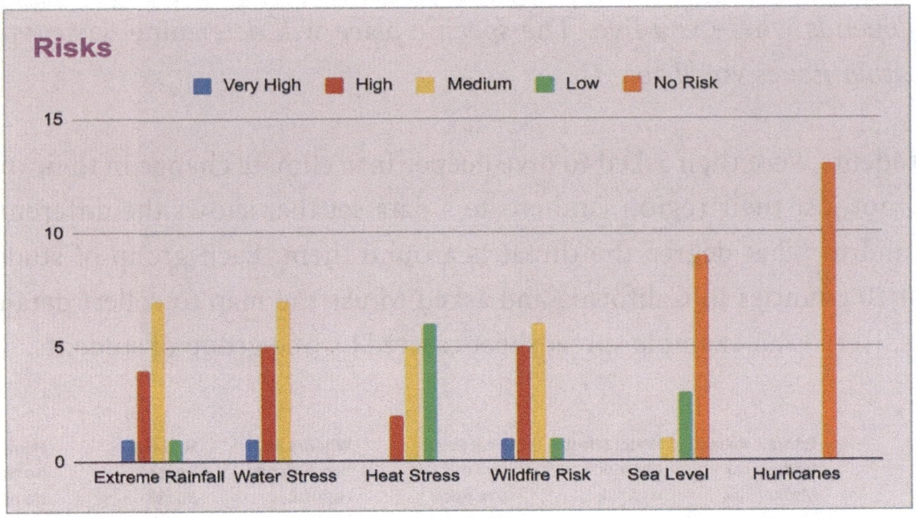

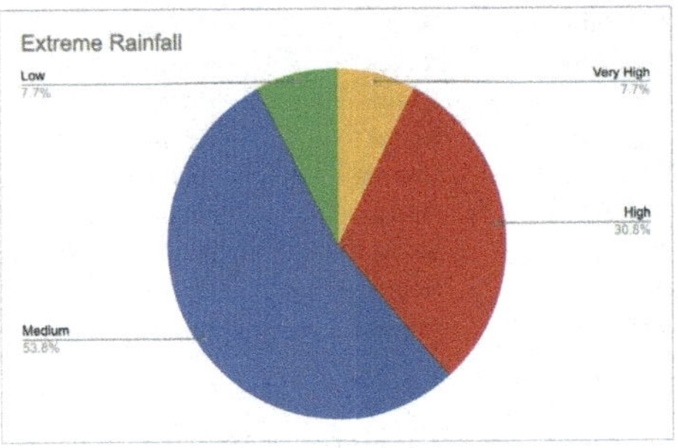

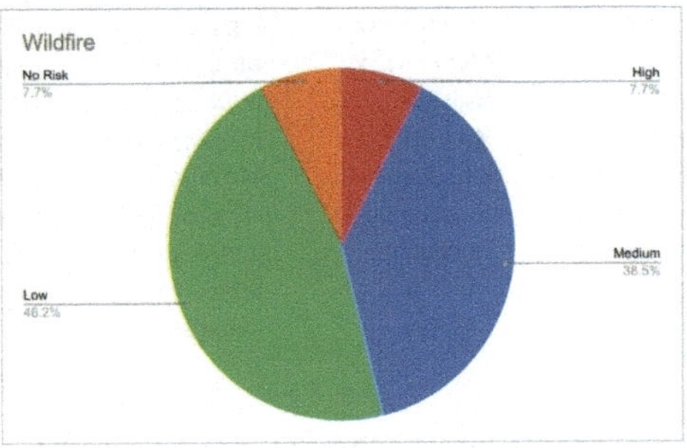

Based on this more local data, students were able to answer the posed investigative question about their local community. This final part of the investigation delves deeper into the data presented in the map beyond the color.

Overall, the goal of this investigation is to use data to deepen understanding and gain more in-depth insight into the topic of climate change. By analyzing the map of the United States, students used multiple perspectives to interpret the data summarized in the map. Students were scaffolded to reflect students' prior knowledge and connection to new information presented on the map. Throughout the investigation, students developed questions that they wanted to investigate based on what they'd observed. They were no longer merely looking at the United States map; now they were investigating the consequences of climate change and asking questions. One student's observation caused him to ask the following question:

> *We wondered why there is one small spot of extreme rainfall in the middle of an extreme drought zone. We want to investigate how that could happen.*

This investigation meets curriculum standards in multiple pathways, and existing current curricula from other subjects can be used in collaboration with this investigation. It is cross-curricular and builds on multiple layers of knowledge, such as language of discipline (science), writing (reflections, information text), reading (theme, big ideas, categorizing and classifying information), and statistics and mathematics (patterns and data talks).

This investigation provides opportunities for students to make connections between old discoveries and new discoveries. They can formulate new theories about climate change, and they can defend those theories using evidence presented on the map. For the sixth grade class in Southern California, through different data, some students generated equations that hypothesized how climate changes occur. Other groups of students connected their findings to previously read articles (e.g., Newsela.com, "The Anti-Greta: YouTuber Campaigns Against 'Climate alarmism') that suggested strong opinions against climate change. Using the data on the map, they generated a rebuttal against that particular article. Overall, students referred to the map as evidence for climate change support.

To culminate the investigation and to relate their data to their own experiences, the sixth grade students were asked to take five pictures of their environment. The pictures were used as part of a photo-voice study conducted by Imogen Herrick and Anna Gralnik[5]. Some examples of student pictures were:

Using the pictures, students can be encouraged to carry out a similar investigation as the one in 2B.1.

5 Herrick, I., and A. Gralnik. Forthcoming. Through the eyes of a child: Empowering and understanding students' climate literacy through pictures. 2021 APA Annual Meeting.

INVESTIGATION SUMMARY:

The main concepts developed in the climate change in our community investigation are:

1. Understanding the context surrounding data is important to be able to draw conclusions from the data.
2. Data can come and be summarized in many different ways.
3. Data can support understanding of large, important issues.
4. Results from data investigations need to be communicated with supporting evidence.

The investigations in this unit are elaborate and can take multiple days to carry out. Students should recognize the role of questioning in working with nontraditional data. Students should also recognize that multiple sources of data and different data sets can be used to answer one investigative question.

References for This Unit

Bargagliotti, A., and A. Gralnik. Forthcoming 2021. Thinking like a data scientist: A cross-disciplinary iInvestigation on climate change. *The Statistics Teacher.*

Bargagliotti, A., Arnold, P., and C. Franklin. 2021. GAISE II: Bringing data into classrooms. *Mathematics Teacher: Learning and Teaching PK–12* 114(6): 424–35.

UNIT 2C:
Exploring Unconventional Data

Unit 2C continues to explore the use of unconventional data. The investigations become more sophisticated and complex, working toward data science. *Data science* is a relatively new term coined to describe statistics in the context of unconventional data. Data science requires thinking about multiple variables at a time (multivariate thinking), asking questions to help sift through larger and more complex data sets, using technology to help wrangle and manipulate data, and understanding appropriate conclusions and limitations to the data. The next several investigations are meant to be used at the high-school level; thus, they rely heavily on technology and students' ability to reason statistically.

Investigation 2C.1: The Trash Campaign[6]

Goals for this investigation: Develop skills for working with multidimensional data and work with apps to visualize data in different ways.

The Mobilize Project (www.mobilizingcs.org/) was a project funded by the National Science Foundation that designed a year-long data science curriculum titled "Introduction to Data Science Curriculum." The goal was for secondary students to develop a blend of computational and statistical skills applied to a variety of data, including Big Data, and in particular data collected in participatory-sensing "campaigns." Participatory sensing (PS) is a data collection paradigm designed to create communities centered on both collecting and analyzing shared data (Burke et al., 2006). The Mobilize project used the term *campaign* to refer to the entire process of collecting data via participatory sensing, including choosing a topic, crafting survey questions, collecting data, and then analyzing and interpreting the data. Participatory-sensing data include many characteristics associated with Big Data, and one goal of the curriculum is to prepare students to reason with data that do not easily fit into a random sampling paradigm. The Trash Campaign is an example of such a participatory-sensing campaign.

6 This investigation was created in conjunction with Rob Gould, principal investigator of the Mobilize Project, and Terri Johnson, a UCLA graduate student in the statistics department.

We begin by creating context and reading brief news articles discussing the Puente Hills landfill, the primary landfill for Los Angeles County. Such articles can be found here: www.npr.org/2014/02/22/280750148/closing-americas-largest-landfill-without-taking-out-the-trash and here www.cnn.com/2012/04/26/us/la-trash-puente-landfill/index.html.

The news articles provide context for the investigation. The Los Angeles County Sanitation Districts (LACSD, www.lacsd.org) would like to reduce their burden on regional landfills, such as the Puente Hills landfill mentioned in the articles. The LACSD is planning a public awareness campaign and wants to ask the public to take specific steps that will help reduce the landfill burden. They would like you to make a recommendation, based on data collected through a participatory-sensing campaign, that would reduce the use of the regional landfills. This task aligns with the following investigative questions:

What steps can be taken to help reduce the landfill burden in Los Angeles?

Students aim to make recommendations for the public awareness campaign supporting their ideas with evidence drawn from data. Data that could be utilized to address the investigative question were collected by the Trash Campaign, as well as by Los Angeles–area high-school biology students and their teachers. These students and teachers recorded data on their mobile devices every time they threw away a trash item over a five-day period. Because the trash data were collected through the Mobilize app based on a trigger—throwing out a piece of trash—they are a participatory-sensing data set. Data collected from multiple classrooms over a one-month period were combined. The students and teachers who collected the data signed waivers to allow for public use of the data, and the data were anonymized by removing names. The Mobilize app can be downloaded for free at www.mobilizingcs.org.

The data collected for the Trash Campaign in LACSD consist of approximately 2,600 observations of 17 variables. The variables are categorical (which type of trash bin the item was placed in; the type of trash item; the activity that generated the trash item; where the trash item was discarded), quantitative (the number of recycling bins visible from the location where the item was discarded; the number of trash bins visible; the number of compost bins visible), image (photos of the trash items), date, time, location (as latitude and longitude), and text (an open-ended description of the trash item).

The set of variables provided and the data collection scheme do not match those of a well-designed, random-sample-based study. Although the investigative question requires making conclusions beyond the sample at hand, the lack of a random sample means that generalizations to a larger population. In general, one would expect student reasoning and analysis to be guided by personal knowledge of recycling and landfills. For example, a student might reason that if more recyclable goods were put in recycling bins, the burden on landfills would decrease. This might lead that student to compute the percentage of recyclable goods that are put into trash bins, compared with another student computing the number of trash bins visible. Although the PS data would be a poor estimate of this percentage for all people in the county, it could serve as evidence of whether a problem, such as the amount of waste Los Angeles produces, does or does not exist.

The data are available in the Trash.csv file. The data can also be accessed through the Mobilize public dashboard at https://sandbox.mobilizingcs.org/#demo/. To access the data from this site, select the Trash Campaign and view it using the dashboard option in the drop-down menu.

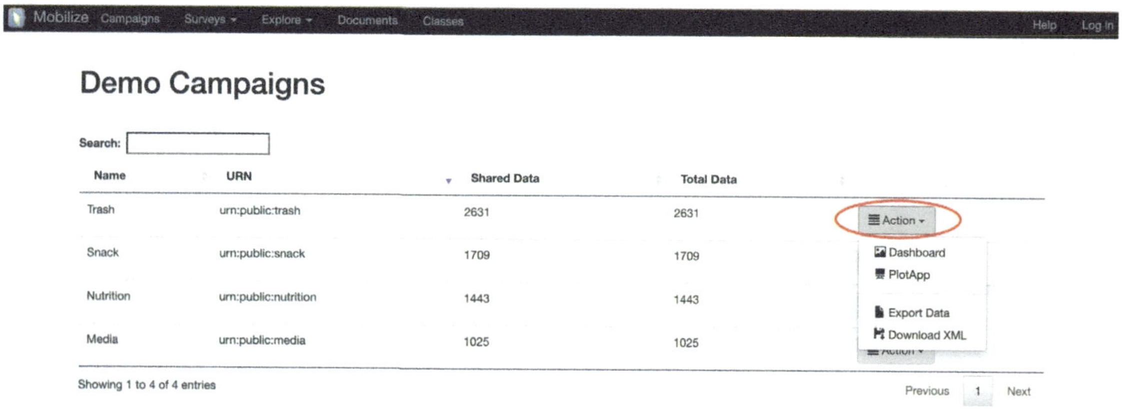

The dashboard provides a limited number of "traditional" visualizations of data. However, it is helpful for exploring multivariate associations in rich data such as these. Below is an image of the dashboard with all of the possible visualizations displayed. Each display has the option of being displayed or not displayed by clicking on the icons at the top of the screen.

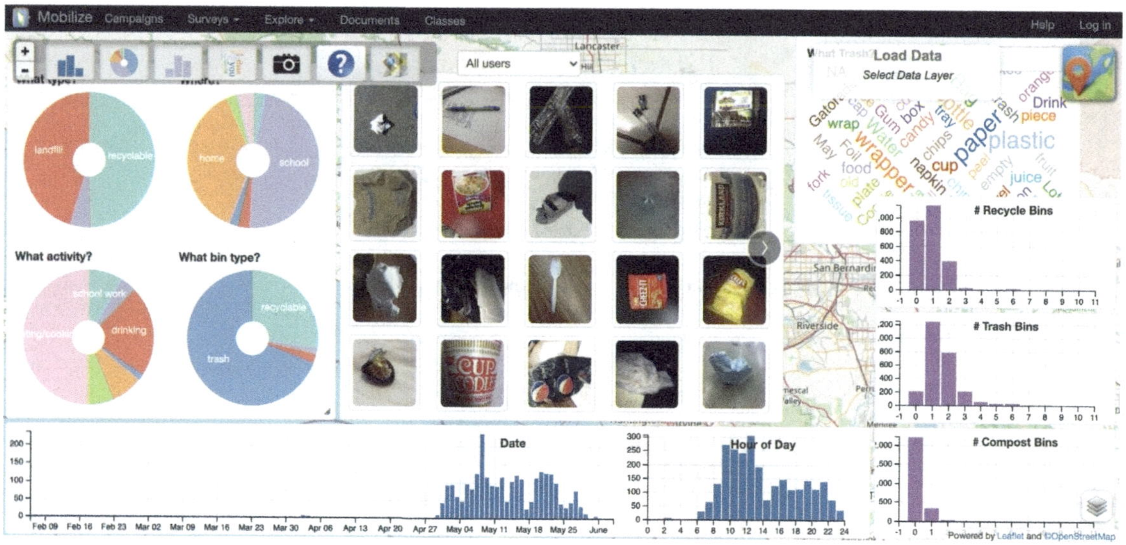

Clicking on almost any part of the dashboard "subsets" the data based on the value clicked and immediately produces a new view of the subsetted data. For example, on the "What activity?" pie graph, we can click on *drinking*, which will make all of the graphs shift to include *only* the trash that was drinking related. The new graphs show only those trash items that were categorized as coming from a drinking activity by the students partaking in the data collection. You can reset the data by clicking *reset*. Clicking on a date in the bar graph titled Date will change the dashboard to display only data from that date, etc. Essentially, clicking on different parts of the pies will illustrate the conditional distributions of all of the variables. Before formally continuing with the investigation, we encourage exploration of the data through the dashboard in order to become familiar with the available variables and visualizations. To help focus this exploration, use the following questions as a guide:

- What activity generated the most trash overall?
- What activity generated the most trash at school?
- Are there differences in the number of observed recycling bins based on the time of day?

To answer the investigative question, it would be beneficial to brainstorm some potential ideas. First, one way the landfill burden could be reduced is if recyclable items were in fact recycled and never thrown in the trash. Similarly, we might be able to reduce the landfill burden if compostable items were composted and not thrown in the trash. Looking to the data, we need to determine whether there are a lot of recyclable and compostable items that are not recycled or composted. To find these data in the dashboard, we first click the *recyclables* portion of the "What type?" pie chart, which provides the following visuals:

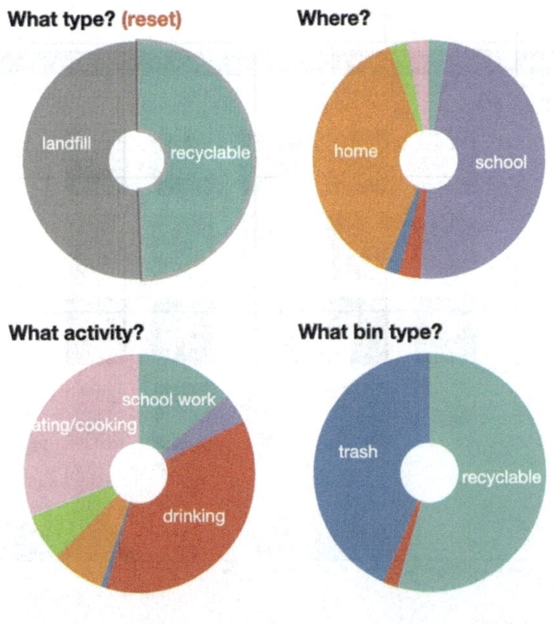

In the "What bin type?" pie graph, we see that many recyclable items are unfortunately thrown in the trash bin. By hovering over the slices of that graph, we can see that 566 items that were recyclable actually ended up in the trash. Conditioning further on that group of 566 items, we can also see that when these items were thrown away, in more than 300 cases, there were no recycling bins in sight. From the word cloud, we can see that these items included things like bottles, paper, and wrappers. Based on these visualizations, one recommendation might be to increase the amount of recycling bins across the city so that recyclable items can be recycled.

We can investigate this potential recommendation further by looking at the location and the activity where these items are being generated and thrown out. The "Where?" pie graph reveals that home and school are the two main locations where these items are being generated. Based on the "What activity?" pie chart, we see that *eating/cooking* is the category that generates the recyclable items that are ending up in the trash. Looking at the "Hour of Day" graphical display, we can see that the items are being largely generated in the morning, from 10 a.m. to 1 p.m. Because these are school hours, it appears that a potentially useful recommendation may be to increase the number of recycling bins in schools. If we make a condition on the location *school* by clicking on *school* in the "Where?" pie chart, we in fact see that many students reported that they did not see any recycling bins when they were throwing out these recyclable items.

Clicking on the picture icon confirms that many of the items that students reported in these categories should have been recycled. The items most frequently seen in the following image are cans, papers, and plastic bottles.

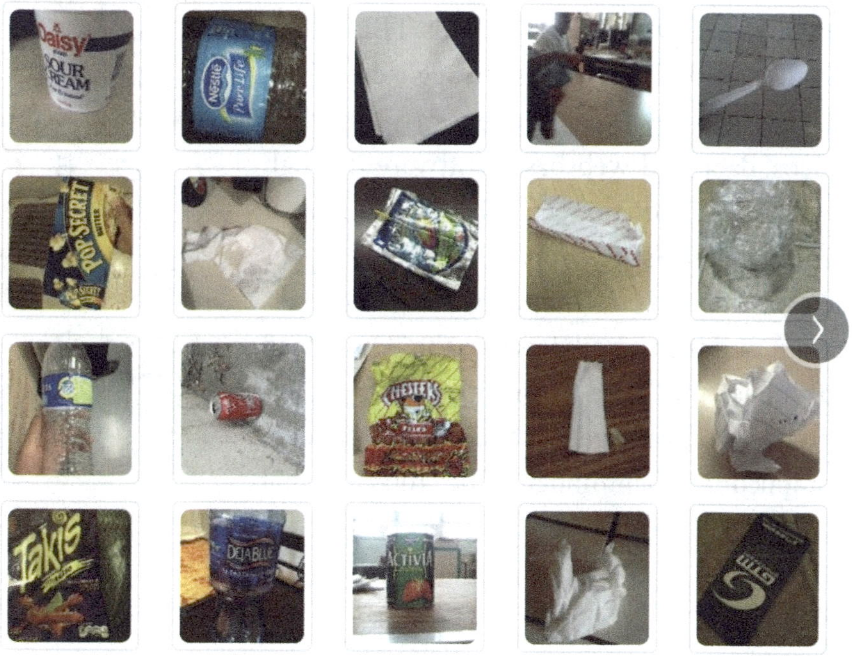

Considering the information gathered in this exploration, one answer to the investigative question would be to recommend that LACSD install more recycling bins in high schools.

While this provides one potential answer to the investigative question, many other recommendations can be explored based on individual interest (e.g., compostable trash, behaviors of students depending on the time of day, etc.). Students should be encouraged to pursue their own ideas to investigate potential recommendations.

Follow-Up Question

1. Complete the trash investigation and offer two other potential recommendations for the LACSD. Write a letter to the LACSD presenting your recommendations using fewer than two pages.

Investigation 2C.2: Gapminder[7]

Goals for this investigation: Develop skills for working with multidimensional data.

The Gapminder Foundation is a Swedish nonprofit foundation dedicated to the achievement of the United Nations' (UN's) Millennium Development Goals through studying statistics on social, economic, and environmental development. Hans Rosling, who passed away in 2017, served as an academic and the chairman of the Gapminder Foundation. Among many other accomplishments, he is well known for his TED Talks, in which he discussed global issues through data. The Gapminder website, www.gapminder.org/, provides access to videos, data, and web-based software through which you can view and interact with global data. If you click on the website's "Resources" tab in the upper right corner of the screen (https://www.gapminder.org/resources/), and then scroll down to the "Download the data" link under the "Data" icon (https://www.gapminder.org/data/) you can see the sources for the data that are compiled in Gapminder. The International

7 This investigation is adapted from https://s3.amazonaws.com/fi-courses/tsdi/unit_1/CheungGapminderlesson.pdf and www.gapminder.org/downloads/teachers-guide-200-years-that-changed-the-world/.

Labour Organization, the World Bank, and the World Health Organization are among the sources. Although the data are collected and recorded in a conventional manner, the data are then compiled and are far more complex, with far more variables, than students and teachers are familiar with at the school level. It is for these reasons that this investigation is included in this unconventional data unit. This investigation serves as an excellent opportunity for students to think multidimensionally and work with graphical displays that illustrate multiple dimensions.

At the turn of the 21st century, the UN put forth its Millennium Development Goals. Information about these objectives can be found on the UN's website: www.un.org/millenniumgoals/. These goals include the following worldwide issues that the United Nations has dedicated its efforts toward improving:

1. Eradicate extreme poverty and hunger
2. Achieve universal primary education
3. Promote gender equality and empower women
4. Reduce child mortality
5. Improve maternal health
6. Combat HIV/AIDS, malaria, and other diseases
7. Ensure environmental sustainability
8. Global partnership for development

To help achieve these goals, nations, organizations, and companies across the globe spend large amounts of money each year to provide aid to regions and countries in need of improvement in each of these eight areas. For example, in 2015, the United Kingdom gave more than $19 billion in economic and military aid to other nations and Germany gave about $16 billion (see www.foreignassistance.gov/explore).

Suppose that a nonprofit has hired you to consult on where it should focus its aid efforts. It has $1 million to donate. It has asked you to focus on one of the UN Millennium Goals and to make a pitch about which area of the world would be most in need of aid. To complete this task, you will choose a UN goal to focus on and recommend the region or country needing support. This task aligns with the following investigative questions:[8]

8 The investigation can be structured in the context of asking students to make a slide presentation of their findings in order to pitch their recommendation to the nonprofit. A maximum number of slides could be allowed (e.g., three slides is sufficient). Slides should present the chosen goal and the recommended region/country for the aid, and be supported by visual displays.

What region or country do you recommend giving aid to in order to support the achievement of your chosen goal? What evidence is there to support your recommendation?

To provide an example of the way this investigation could be approached, we choose to work with the Millennium Goal of reducing child mortality. To begin, click on the Gapminder "Tools" tab, also on the "Resources" page. This tab will load the data set in a scatterplot that has bubbles of different colors and sizes instead of points. There are four dimensions pictured in this plot: The x-axis defaults to *income per person,* the y-axis defaults to *life expectancy,* the colors represent which region of the world the countries correspond to, and the size of the bubbles show the population size of the country. Each bubble on the scatterplot represents a country. A user can change three out of the four variables being represented in the visual by hovering over the names of the variables on the axes and using the drop-down menu to choose the variable to represent the size of the bubble.

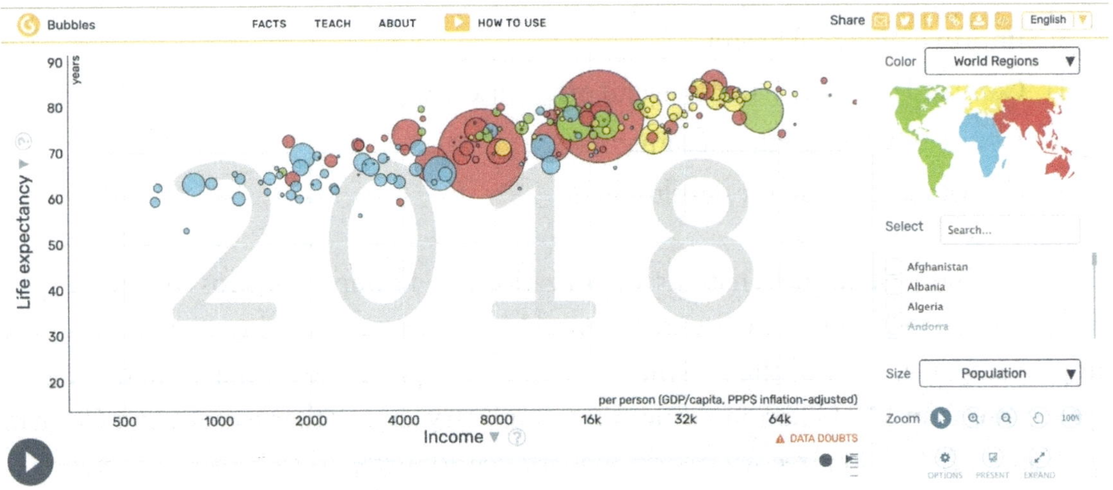

To address our chosen Millennium Goal, we first have to visualize the variables that pertain to it. To change the variables represented on the graph, click on the y-axis variable *life expectancy,* and a drop-down menu of variables will appear that we can choose from. When doing this activity with students, teachers are encouraged to choose the variables for their students to explore beforehand. Because of the large number of variables available in the data set, students may get lost trying to decide what to look at. Therefore, if a teacher chooses a list of, for example, five initial variables for students to explore, students would have the opportunity to investigate the data set, which keeps the investigation student-led, but in a controlled and scaffolded manner. Depending on the level of students, an instructor could allow students to choose their own variables to investigate. For example, the instructor could ask students to choose three variables to investigate that relate to their chosen Millennium Goal.

For our chosen goal of reducing child mortality, we begin by looking at the following variables:

- Child-mortality rate
- Babies per woman
- Income per person

We hypothesize that richer countries might have lower child-mortality rates. We also hypothesize that the more babies women have, the higher the child-mortality rate. Thus, we pose the following question to help guide our analyses:

To what extent do countries that are richer have lower child mortality rates?

Plotting *income* against *child mortality*, we see a downward linear trend, thus illustrating a negative association between income per capita and child-mortality rates. Playing the video shows how this relationship progresses through the years; we see that initially. In 1800, all countries were closer in income, and they all had high child-mortality rates, at around 500 deaths per 1000 births. Toward the beginning of the 1900s, we can see the European countries (those in yellow) and the United States pulling away from the others in terms of wealth. The red dots of New Zealand and Australia are also in the mix of wealthy countries, as well as South Africa, which is represented by a blue dot. By 1950, almost all of the European countries, the United States, Australia, and New Zealand had drastically cut their child mortality rates to below 80 per 1000 births. In addition, these countries had also pulled ahead of the others in terms of income per capita. As we continue to 2018, the final year of data available for these two variables, we see that every country has reduced its child-mortality rates. However, for the most part, the countries with the lowest income and highest child-mortality rates are the blue countries, while Canada, Europe, and red countries such as Japan and South Korea have high income and low child-mortality rates. The United States trails a little behind, with high income and slightly higher child-mortality rates. These progressions suggest an association between a nation's wealth and the number of child deaths in that nation. From these progressions, we can also observe that the region of the world most in need of support to reduce child mortality is Africa.

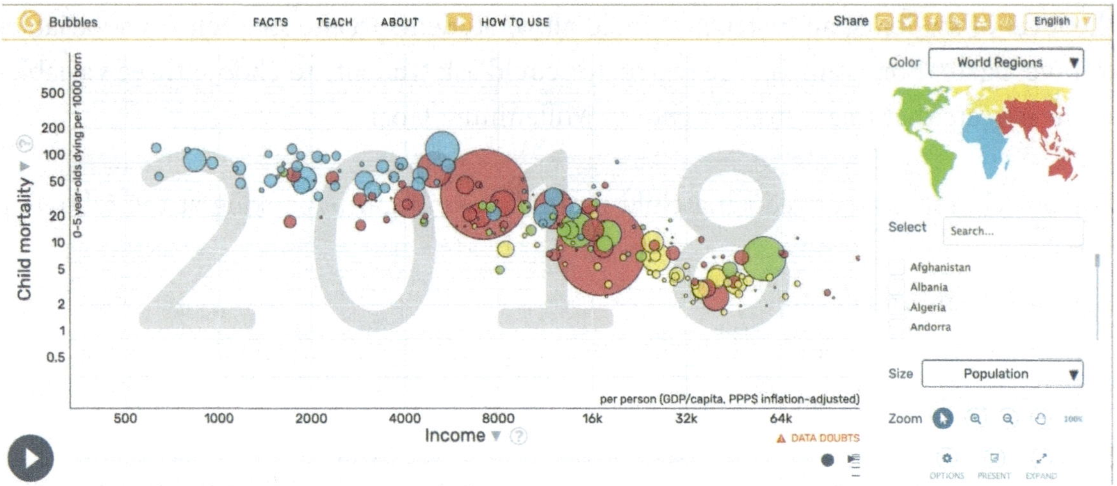

Income Versus Child-Mortality Rate

Because we have identified a region where the aid could be given, students might think the investigation is complete. However, it is unclear what factors may be contributing to Africa having a large child-mortality rate. To allocate aid properly, we need to investigate further the potential reasons Africa might be having difficulty, therefore giving the non-profit organization more direction to where to designate its aid. We now try to further unpack potential reasons for high child-mortality rates by examining:

To what extent does the number of babies a woman gives birth to affect child-mortality rates?

To look at this relationship, we change the x-axis variable to be the number of babies per woman. By looking at the 2015 scatterplot of this relationship, we see a positive trend, with African countries producing the highest number of children per woman. Recall our previous analysis regarding the high child-mortality rates of the blue countries. By adjusting the bubble size according to the income of the country instead of the population, we see that some of the blue countries that have high child-mortality rates and high numbers of babies per woman are wealthier than others. For example, Guinea, Gabon, Angola, and Nigeria have larger incomes than most of the other African nations. When we view the time video, we see that all countries started with a high number of babies per woman; however, as time progressed, other countries started having fewer children, while the African countries remained at the high birth rates of approximately five to seven children per woman.

This may make us curious as to why African countries have not decreased the number of children per woman over time, so we explore the answer to the following question:

To what extent is contraception available in countries around the world?
How is access to contraception linked to the number of babies a woman has?

Plotting contraception use against time, we see that for the most part, African nations have a very small percentage of people using contraception (approximately 20%). Although a few African countries do have higher rates of contraception, the majority of the countries have low rates. When plotting contraception rates against the number of babies per woman, we see a linear relationship between these two variables. As contraception rates decrease, the number of babies per woman increases.

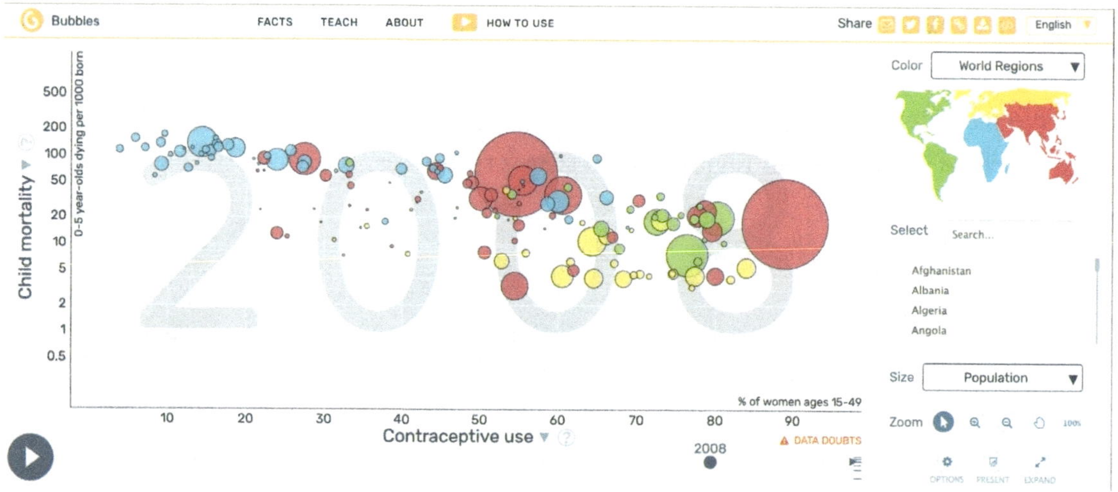

All of these explorations suggest that a possible way to support African countries with their child-mortality rates is to reduce the number of babies a woman has. This can be done through increasing access to contraceptives. However, this plan does not appear to get at the root of the problem of *why* children in African countries are dying at higher rates than anywhere else. Is it simply that there are more children being born in Africa overall? Or can we dig deeper and try to recommend something more specific to explain why more deaths might be occurring?

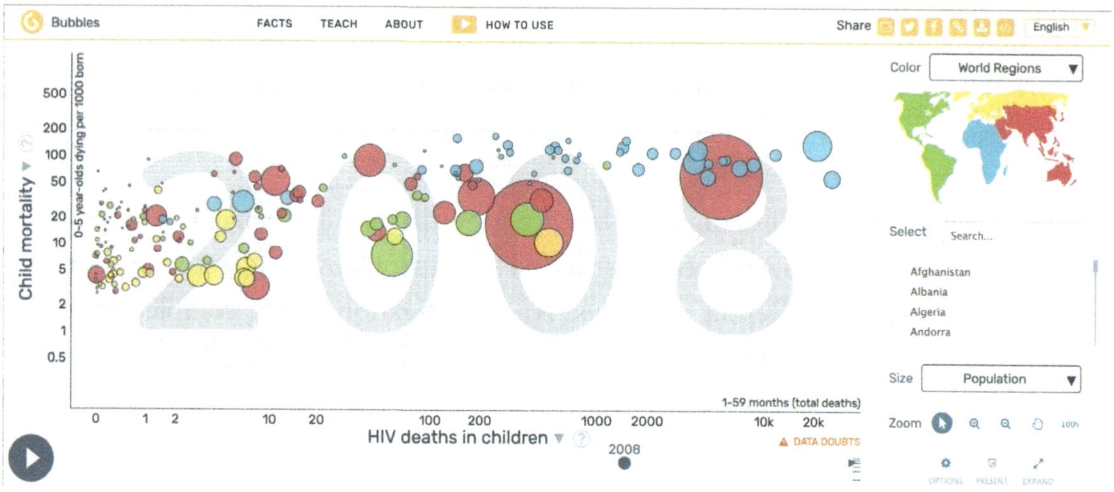

We can examine links between child mortality and two other variables: HIV rates and malaria rates. To examine these links, plot the child-mortality rate with the rate of HIV deaths for children to see that HIV in children is greatly affecting the African region. Although South Africa and Eswatini (formerly Swaziland) are two of the wealthiest African nations, they also have very high HIV rates. Looking at malaria rates, we also see that this is another large cause of death among children in Africa. Although the country of Guinea is wealthier than other African countries, it still has high rates of malaria. The Doctors Without Borders website (www.doctorswithoutborders.org/) discusses the hardship of malaria exposure in Africa. This organization states that although there is an effective treatment for malaria, called artemisinin-based combination therapy, there is a large shortage in the African region due to scarce resources.

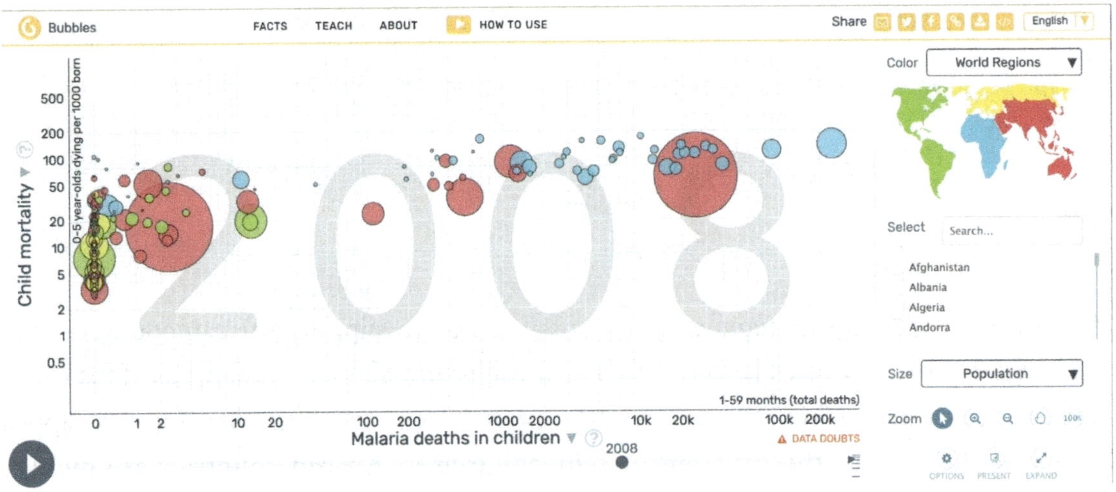

These findings suggest some targeted recommendations:

- Focus on child-mortality rates in the African region, because it is the largest need within the chosen Millennium Goal of reducing child mortality
- Increase contraception availability and usage in African nations to reduce the number of babies per woman and the spread of HIV
- Increase treatment and availability of treatment for malaria in African nations

Although we began this investigation by targeting three initial variables to examine, our investigation led us toward other variables, in a way uncovering a story as our curiosity led us to look at different variables. An investigation is not complete until the message one is trying to convey is well understood and, through exploration, one is able to offer concrete recommendations based on the data. Hans Rosling's Ted Talk (www.gapminder.org/videos/reducing-child-mortality-a-moral-and-environmental-imperative/) discusses child-mortality rates around the world and how the data on child mortality are collected.

It is important to note that this investigation could go down many different paths. The connections students explore might be different from the ones described. While working with students on an investigation of this type, it might be tempting to quantify the number of connections students need to make or the number of variables students need to look at. However, in statistics, particularly when using data sets of this sort, the different pathways a student could take are numerous. Students should be encouraged to draw connections until their recommendations are concrete and supported by the data.

Additionally, the Gapminder website has multiple videos on how teachers have used the site in their classrooms, accessible through /www.gapminder.org/for-teachers/.

Follow-Up Questions

1. Conduct the Gapminder investigation for each of the UN goals.

Investigation 2C.3: Global Terrorism and Religion[9]

Goals for this investigation: Develop skills for working with multidimensional data and work with apps to visualize data in different ways.

Terrorism has always been a concern for modern society. In this investigation, we will use the Global Terrorism Database (GTD, www.start.umd.edu/gtd/) to examine the relationship among a particular region of the world, religion, and terrorism. The GTDis an open-source database that contains information about more than 150,000 terrorist incidents occurring between 1970 and 2019. The data in the GTD are gathered from news reports, and the team managing the database tries to verify all the information it gathers through multiple news sources to make the database as reliable as possible.

Using GTD, we aim to answer the following question:

How do religion and region of the world affect the presence of terrorism?

We will access the GTD data through the Grinnell College RStudio server, managed by Dr. Shonda Kuiper, at http://shiny.grinnell.edu/. One can also go directly to https://shiny.grinnell.edu/GlobalTerrorismMap/ and https://shiny.grinnell.edu/GlobalTerrorismPlots/ to access the data. Information about the data can be found at https://stat2labs.sites.grinnell.edu/GlobalTerrorism.html. The data are presented in the form of a world

9 This investigation was created in conjunction with Shonda Kuiper. See https://stat2labs.sites.grinnell.edu for Kuiper's extensive contributions to statistics and data science education.

map. A map of this type can be useful for looking at patterns over time, picking out individual incidents, and getting detailed information. You will see a map of countries, overlaid by incident markers of different sizes. The size of each marker represents the severity of the attack (a weighted sum of the deaths and injuries caused by the incident). From the drop-down menu in the top corner, you can select the region of the world you are interested in looking at. The map will shift to that region, and you will see dots on the location of the terrorist attack.

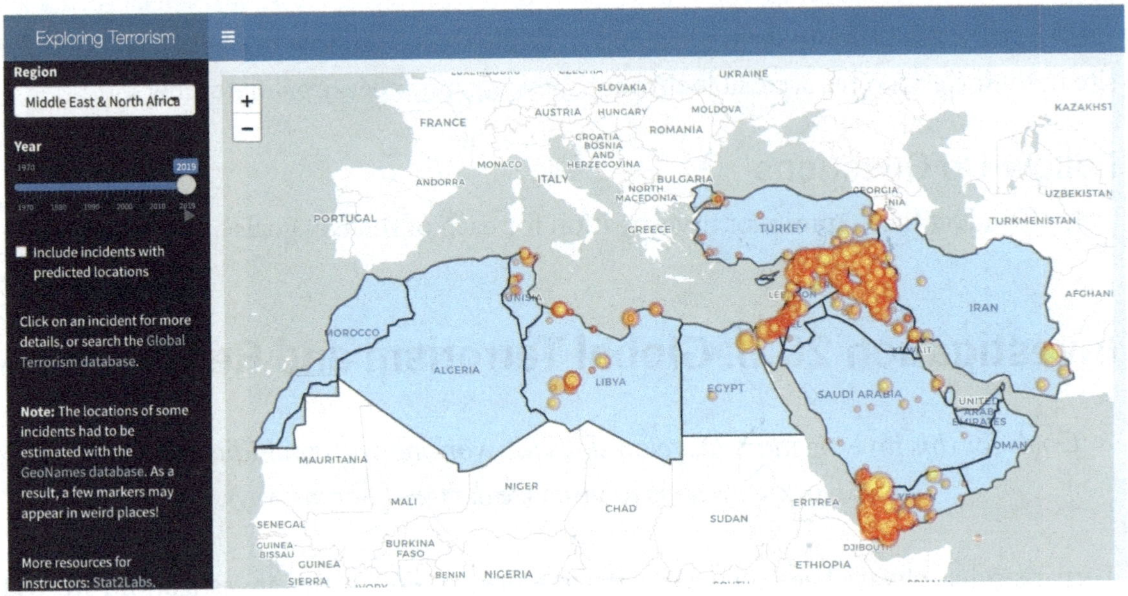

As a first pass at trying to understand the data, we select each of the different regions one by one and press play on the timeline to see the terrorist acts unfolding over time from 1970 to 2014 in each region. Doing this reveals the following overall patterns:

- Terrorism has plagued the world throughout all years. There appear to be very few "quiet" years across the globe.
- Terrorism seems to be more prominent in specific regions at different historical time periods:
    - The Middle East and North Africa had large amounts of terrorist attacks during the early 1980s and '90s, with Israel, Palestine, and Algeria having most of the attacks in the region.
    - North America has had a consistent amount of terrorist attacks through the years, with a few surges in the mid-to-late '90s and in 2001. Overall, the attacks in North America have not been severe.
    - South Asia had some attacks in the '90s and numerous attacks in the mid-to-late 2000s.
    - Sub-Saharan Africa has seen numerous attacks since the '80s. There have been a large number of attacks in Nigeria, Sudan, Somalia, and Uganda, and

while other African countries have also had terrorist attacks, these countries appear to have had the most severe ones over the years.

  - Europe and Central Asia have had numerous attacks, particularly in the '70s and '80s, with a decrease in the 2000s.
  - Terrorist attacks in Latin America and the Carribean were prominent in the '80s and '90s, particularly in Guatemala, Colombia, and Peru. In most countries in the region, the attacks had subsided by the 2000s, with the exception of Colombia, which had many attacks throughout the 2000s.
  - East Asia and the Pacific have some specific countries that have seen many attacks throughout the years, and others that have had hardly any. The Philippines stand out as the country in the region with a consistent amount of attacks throughout this time period.

These observations summarize the overall patterns of terrorism across the different regions of the world since 1970. Next, we examine the link to religion. Looking at external sources, such as https://ca.pbslearningmedia.org/resource/sj14-soc-religmap/world-religions-map/, we find that the major world religions are:

- Christianity (32%),
- Islam (23%),
- Hinduism (15%),
- Buddhism (7%),
- Sikhism (.4%), and
- Judaism (.3%).

Using the plots app, we can select variables to plot and color the plots using multiple variables. In addition, we can change the time increments displayed. By clicking the "Filters" tab, we can filter the data further by region, type of attack target, and the type of weapon that was used in the attack. Similar to investigation 2C.2, these plots can display multiple dimensions in one plot. On the top bar, we have three options for the type of plots: scatterplot, stacked-line plot, and bar chart. We can make use of all of these plots.

Have terrorist attacks been predominantly present in countries with specific religions? To look at the link among religion, region, and terrorism, we begin by using a stacked-line plot. Select the following graphical display:

- Y-axis: incidents
- By count
- Color by: religion

This shows the following plot.

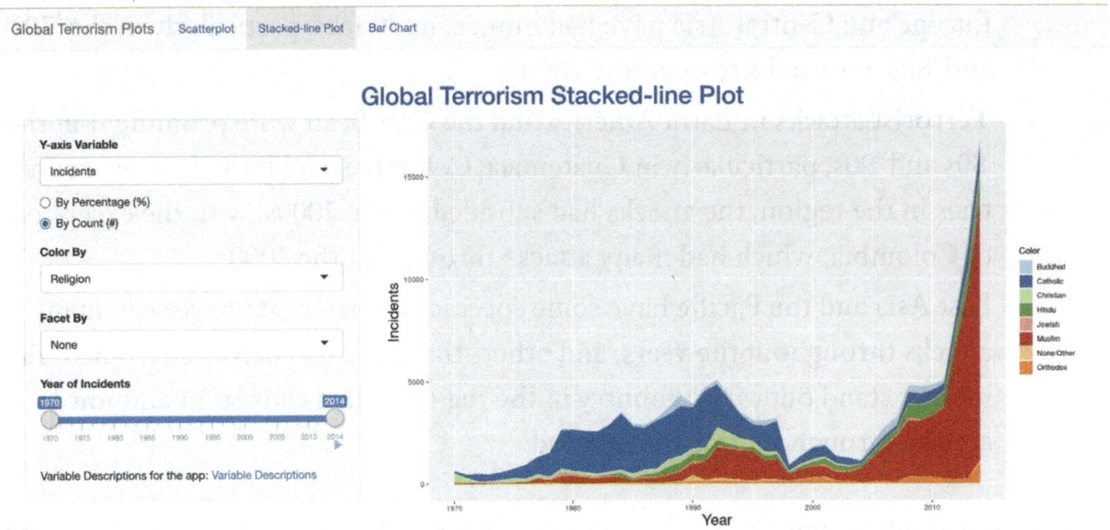

The plot illustrates that in the '90s, terrorist incidents were happening in predominantly Catholic countries, whereas starting around 2007, they were happening mostly in Muslim countries. The plot also illustrates that since 2010, the number of terrorist attacks has spiked dramatically compared with historical data.

Where are these attacks taking place? Have the attacks been concentrated in specific parts of the world? To investigate the location of where these attacks are taking place, we can "facet" the data by region and obtain the following plot:

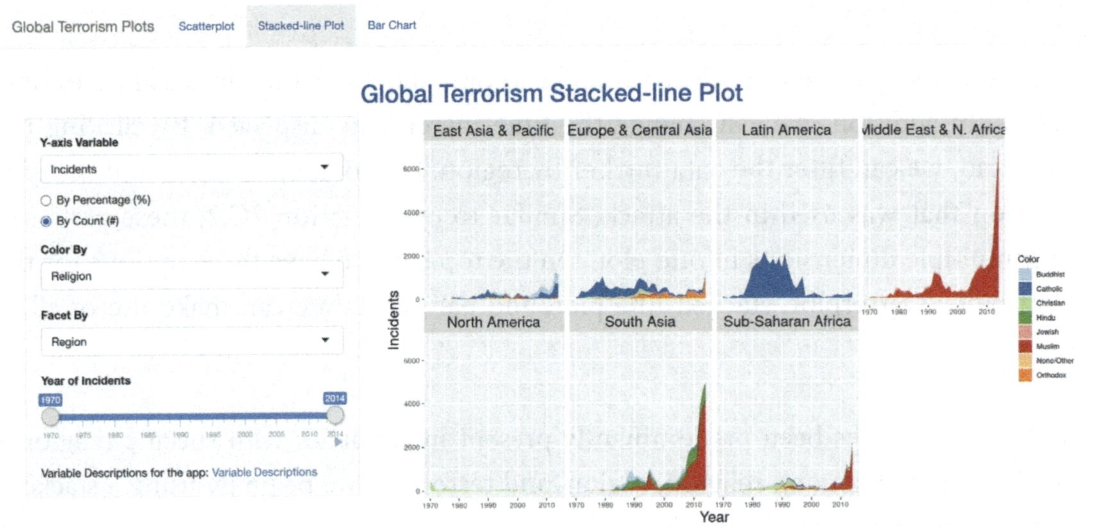

We see that the attacks carried out in predominantly Catholic countries in the '90s were mostly happening in Latin America, and the attacks happening in predominantly Muslim countries in the 2000s were happening in the Middle East, North Africa, and South Asia.

To summarize our findings so far:

- Terrorist attacks have been consistent since 1970, with a large spike since 2010
- These attacks have been taking place in predominantly Catholic and Muslim countries
- The locations of the attacks are mostly in the Middle East, North Africa, South Asia, and Latin America

Are all countries within these regions experiencing terrorist attacks that are leading to fatalities, or are only specific countries susceptible? To investigate this question, we use the bar chart option. Select the following:

- Y-axis variable: fatalities
- Type of range: below/above *n* fatalities
- Value for *n*: 0
- Year of incidents: try 2014

This display illustrates that although there were numerous attacks across the globe, in a large majority of the countries, the attacks led to no fatalities. In fact, in 2014, about two-thirds of the countries experienced no fatalities, while one-third did. The attacks that led to more than one fatality tended to be in Sub-Saharan Africa, the Middle East, and North Africa.

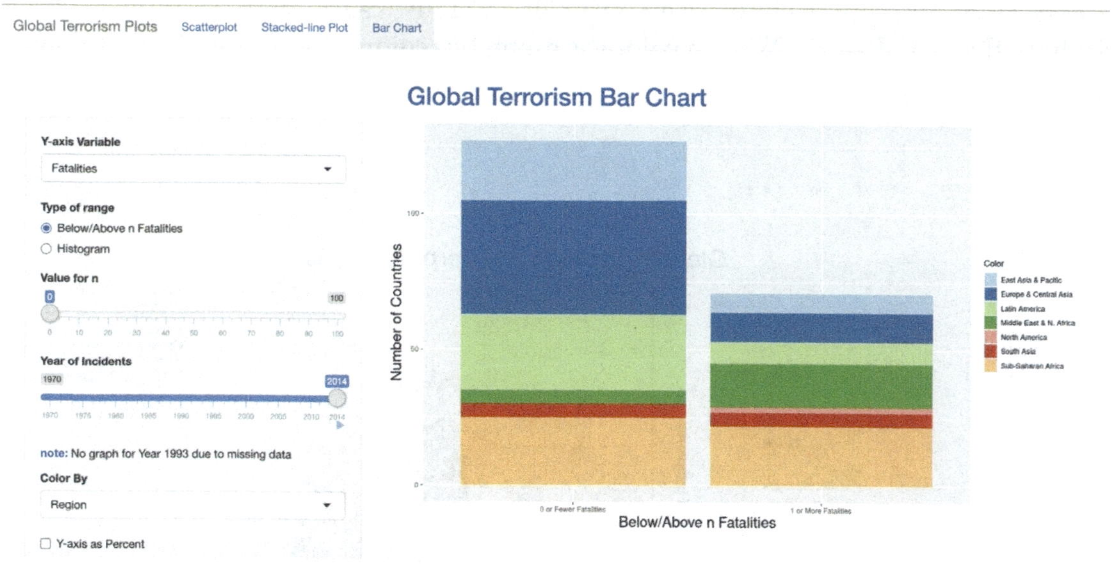

Creating the same display, but colored by religion, we see that the attacks with more than one fatality were primarily in Muslim countries and in Catholic countries.

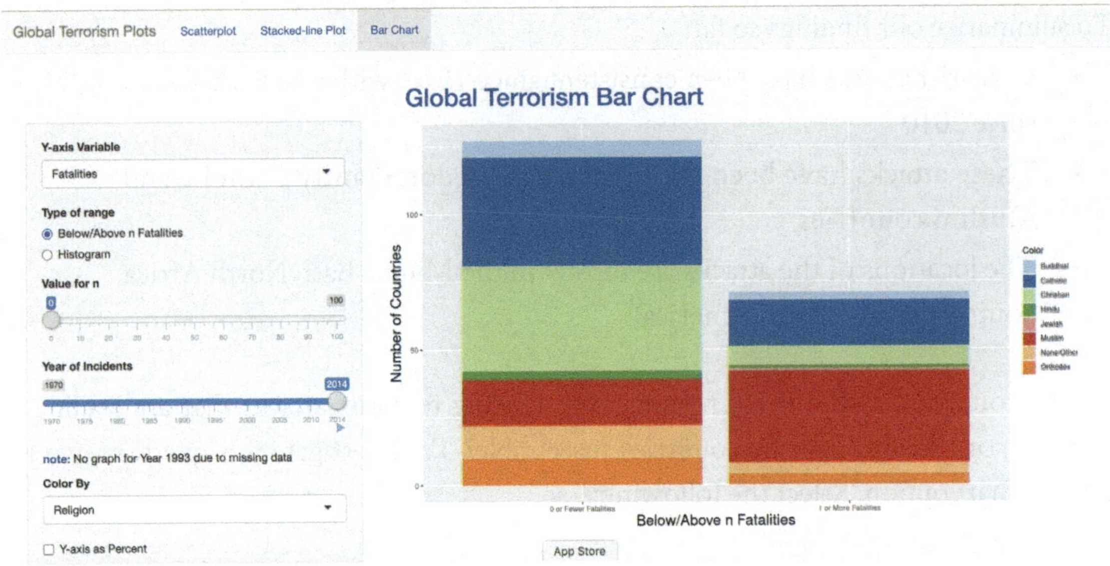

We can now ask ourselves if there are economic factors that might affect whether a region or country has terrorist activity. It can be hypothesized that underdeveloped or high-poverty areas are more susceptible to terrorism.

Are there economic factors that contribute to a region's susceptibility to terrorist activity? To investigate this question, we turn to the scatterplot.

Keeping incidents on the y-axis, we can select the different economic variables for the x-axis. To see the patterns more clearly, we can select the logarithm of the variable option for both the y- and x-axis. We can color the graph by region. We start with gross domestic product (GDP).

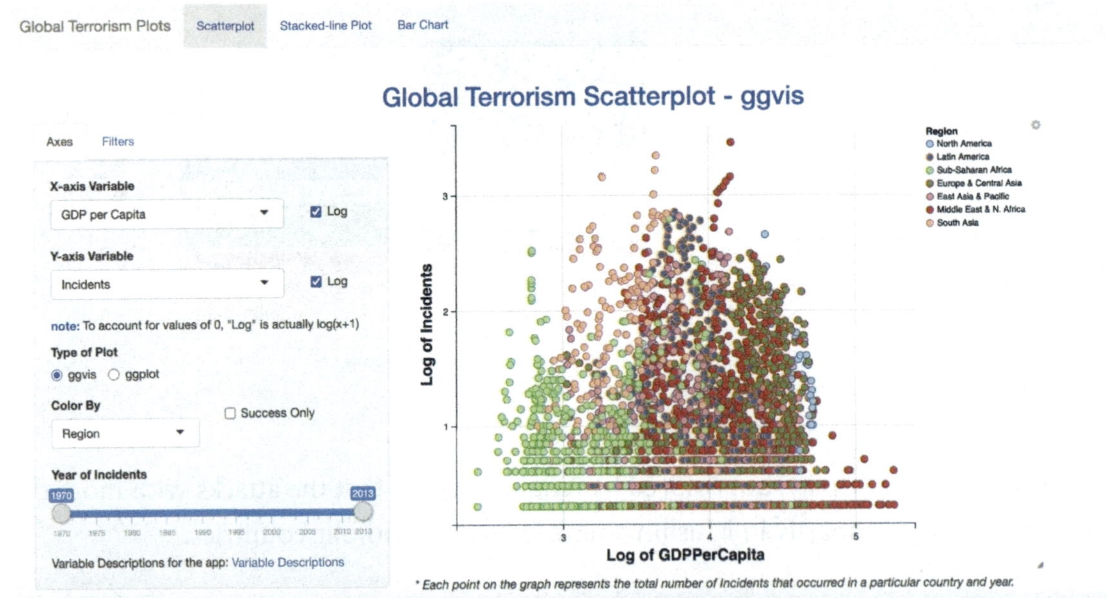

We can see that although there are regional discrepancies in GDP (the green dots for Africa have lower GDP overall than, for example, the blue dots representing North America), regardless of GDP, all regions have had attacks. There does not seem to be any overall pattern relating GDP with the number of incidents. Similarly, when we examine the relationship between the other variables, such as life expectancy, unemployment rate, etc., and number of incidents, we see no relationship. This suggests that there are no clear economic factors that affect the number of terrorist incidents.

Our results also show that terrorism is not random and sporadic. Instead, the distribution of terrorism attacks has been clustered around areas that have religious and ethnic conflicts. While countries that are predominantly Catholic and Muslim have seen the most terrorist attacks, this does not hold true across all of the years analyzed. Catholic countries were experiencing terrorism in the '80s and '90s, while Muslim countries were experiencing terrorism in the 2000s and 2010s.

We created multiple graphs to identify patterns with terrorism and the major religion of a country. With this method, we found that some religions do correlate with greater levels of terrorism, particularly Catholicism and Islam. However, this connection applies only for certain time periods (Catholicism during the '80s and '90s and Islam during the 2000s and 2010s). However, we cannot claim a causal relationship between terrorism and religion. In fact, each year the GTD shows that there are numerous Muslim and Catholic countries with no recorded terrorism incidents.

These data were difficult to navigate for multiple reasons. First, the nature of the questions are open-ended, and there are no clear-cut answers. The investigation is exploratory, and the connections we are trying to make are not causal but are merely exploratory. Second, in this investigation we are navigating several interactive apps—the map as well as the three different graphical plots. Each one of these tools facilitates the analysis of different portions of the data. To answer the investigative questions, all of these parts need to be put together. Furthermore, while the data and the apps are revealing, there are limitations. For example, the countries are coded according to the most prominent religion present in the country; however, many countries, such as the United States, have a mixture of religions represented within it. Thus, we have no way to control for prevalence of religion in this investigation. Also, while many terrorist groups use religion as a motivation, we cannot say that many religious people are terrorists.

This investigation illustrates the difficulties with navigating complex data sets. Using interactive apps helps visualize multivariate relationships among variables; however, it requires practice and time. Because of the large amount of information that can be accessed, approaching the analysis through questioning is particularly important and guides the investigative process.

Follow-Up Questions

1. Using the apps, answer the following investigative questions:
   - What are the predominant types of attacks? Are the methods of attack sophisticated (e.g., hijackings and facility attacks) or not (e.g., assault and bombings)?
   - What type of weaponry is used?
   - Have these patterns stayed consistent over time or have there been shifts in terrorism methods over time?
   - What countries have seen the most fatalities from terrorist attacks in the 2000s and in the 1990s?
   - Who have been the targets of the terrorist attacks?

Case Study 5: Fitbit Tracking

In January 2018, several news sources reported on the unintended consequence of Strava Labs publishing a heat map of people going jogging around the world, using their Fitbit devices as trackers. Doing this inadvertently alerted people worldwide of where secret military bases were located. These Fitbit data are a type of unconventional data that are ever changing and growing. The visualization of the data as a heat map is also an increasingly common way to illustrate these data. Several articles were published on the topic, including the following:

- *The Washington Post*:
  www.washingtonpost.com/world/a-map-showing-the-users-of-fitness-devices-lets-the-world-see-where-us-soldiers-are-and-what-they-are-doing/2018/01/28/86915662-0441-11e8-aa61-f3391373867e_story.html?utm_term=.37ed891587f8
- *Wired*:
  www.wired.com/story/strava-heat-map-military-bases-fitness-trackers-privacy/
- BBC News:
  www.bbc.com/news/technology-42853072
- *The Guardian*:
  www.theguardian.com/world/2018/jan/28/fitness-tracking-app-gives-away-location-of-secret-us-army-bases

The articles all highlight the fact that unconventional data are in the news. While the discussion at the beginning of this unit emphasized how these types of unconventional data do not set us up for inference, here is a case where people were able to infer locations from unconventional data. Although this is not a classic statistical-inference question, data of these types do offer large amounts of information. The articles note that the data offer a large amount of information to anyone who wants to attack or ambush U.S. troops in or around the bases, as well as patterns of activity inside the bases. The *Post* article states, "Many people wear their fitness trackers all day to measure their total step counts, and soldiers appear to be no exception, meaning the maps reveal far more than just their exercise habits." From a data perspective, this means that data are not only collecting location, but they are also collecting time of day, amount of activity, and other related pieces of information. Such fitness devices also showed routes in and out of bases, time of activities in and out of bases, and soldiers' overall daily patterns.

This case study demonstrates how unconventional data now play a role in our daily lives. Along with this come issues of privacy, sensitivity, and implications. These are discussions that should be forthcoming in our statistics curriculum as we fine-tune our understanding of how these types of data factor into society. One thing is certainly clear: Our students have access to, contribute to, and are well aware of these types of data, so the time for engagement with these data in a classroom setting is upon us.

References

Burke, J.A., Estrin, D., Hansen, M., Parker, A., Ramanathan, N., Reddy, S., and M.B. Srivastava. 2006. Participatory sensing. Workshop on World-Sensor-Web (WSW'06): Mobile Device Centric Sensor Networks and Applications.

Cheung, L. 2015. Gapminder: Investigating world issues. In *Teaching statistics through data investigations MOOC-Ed,* Friday Institute for Educational Innovation: NC State University, Raleigh, NC. Retrieved from https://s3.amazonaws.com/fi-courses/tsdi/unit_1/CheungGapminderlesson.pdf.

UNIT 3: Probability Unpacked

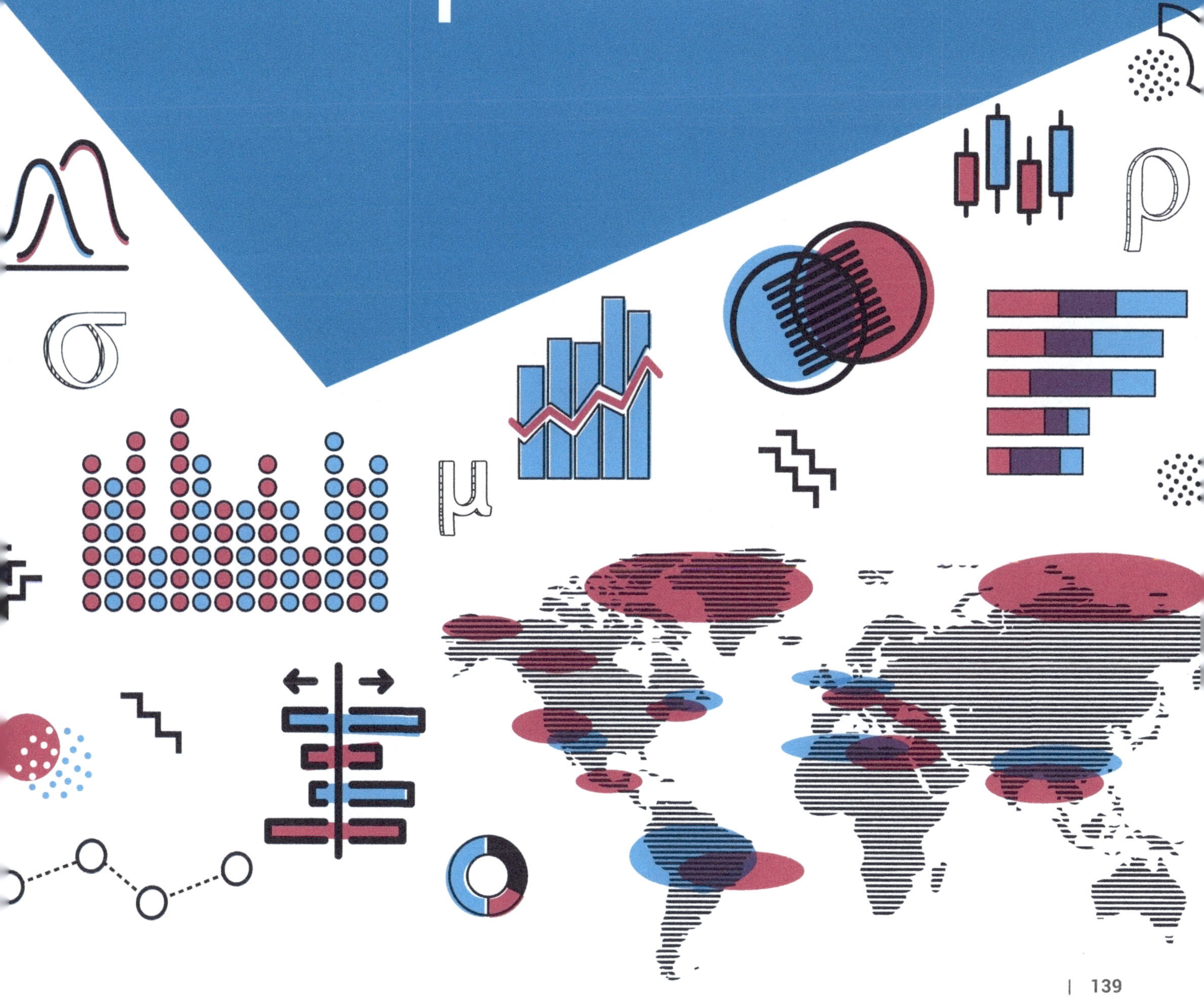

UNIT 3A: Probability Introduction

Probability topics are sometimes taught in conjunction with or before statistics topics in the school curriculum. For example, several curriculum standards place probability and statistics into one strand (e.g., the NCTM's Principles and Standards for School Mathematics, Common Core State Standards). While probability and statistics are related, it is often difficult to see the explicit connection between the two subjects. Typically, probability is introduced using counting rules. These rules allow us to count the number of various outcomes, such as the number of possible sandwich combinations at a restaurant, the number of different gender combinations of babies a woman could give birth to, or the number of possible pathways from point A to point B on a city map. These counting exercises are typically followed by teaching probability rules, such as finding the probability of rolling a die and getting a six or flipping a coin and getting heads. But how is the study of these rules related to statistics?

The answer is randomness. Statistics aims to draw conclusions in the presence of variability in data. Introducing randomness as part of data collection allows us to use probability to develop models for describing the resulting random variation present in data.

As noted on page 11 in *GAISE II*, "Probability is also used in statistics through randomization—random sampling and random assignment. Samples can be collected at random and experiments can be designed by randomly assigning individuals to different treatments. Randomization minimizes bias in selections and assignments. It also leads to random chance in outcomes that can be described with probability models." (The idea of random selection and random assignment will be developed in Unit 3B.)

Probability gives us the tools to model and quantify that randomness. **A probability is a quantity between zero and one that defines how likely something is.** For certain types of random processes (such as tossing a coin or rolling a die), the probability represents the "long-run" relative frequency of an event. In other cases, such as sporting events, the probability represents the likelihood that something will occur.

The idea of randomness begins with the notion that an individual outcome from a repeatable random process cannot be predicted with certainty. However, if the random process is repeated a large number of times, a predictable pattern in the relative frequency of outcomes generated from this process will emerge. Probability is the branch of mathematics that seeks models for describing this long-run predictable pattern. These models provide order to the seeming disorder present in the outcomes from the individual trials.

A **random process** is a process for which an individual outcome is unpredictable. A simple example of a random process could be flipping a coin. When you flip a coin, you do not know with certainty what the outcome of that flip is going to be; it could be heads up, or it could be tails up. In fact, you do not know exactly what proportion of heads will be seen in the first 10 flips. However, assuming a fair coin (each side of the coin has the same chance of occurring), we do know that if we flip a coin a large number of times, in the long run about half of the flips should land heads up and about half should land tails up. In the short run, outcomes are unpredictable, but in the long run, there are patterns. While the individual outcome (e.g., one coin flip) or short-term sequences of outcomes (e.g., 10 coin flips) of a random process are highly variable, making predictability difficult, if the process is repeated a large number of times, then predictable patterns will emerge in the outcomes. The graph illustrates the long-run relative frequency of heads for a fair coin and how it tends to stabilize around half as the number of coin flips increases toward 1000.

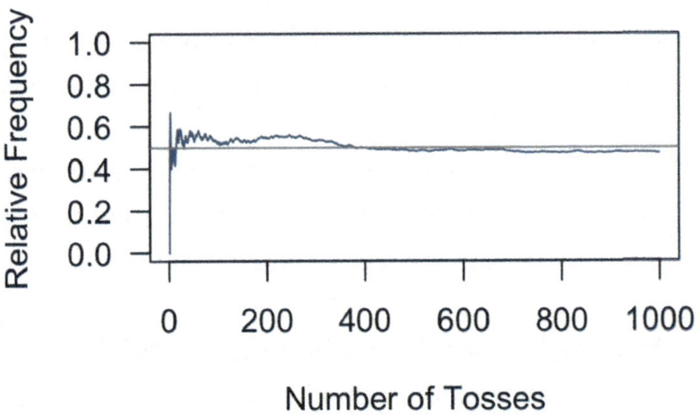

If the coin is assumed to be fair, then a mathematical model for the random process of tossing a coin is:

$$\text{probability of heads} = P(\text{heads}) = \frac{1}{2}$$

Note that this model does not imply that after an even number of trials, a head will occur exactly half of the time. For instance, after two tosses, you are not guaranteed to have exactly one head, and after 100 tosses, you are not guaranteed to have exactly 50 heads. See, for example, the following table, where after 1000 tosses, there were not exactly 500 heads.

Now suppose you are not willing to assume a coin is fair. How could you go about estimating a probability model for this situation? To estimate the probability, you could toss the coin a large number of times and use the long-run relative frequency of heads to *estimate* the probability of heads. For example, the following table illustrates the outcomes:

Trial Number	Outcome	Cumulative Frequency of Heads	Relative Frequency of Heads
1	T	0	0/1 = 0.0
2	H	1	1/2 = 0.5
...	...	...	
1000	H	475	475/1000 = 0.475

After 1000 tosses of the coin, a head has occurred 475 times. Based on these 1000 tosses, we would estimate the probability of a head to be 0.475. This strategy for estimating and assigning probabilities is called **empirical probability**. Based on these empirical probabilities in the table, the coin does not appear to be fair.

The **empirical probability** of an outcome can be defined as the long-run relative frequency of the outcome. The **law of large numbers** states that when performing the same random process a large number of times, the average result obtained in the large number of trials is close to the actual probability. In probability, this indicates that as a random process is repeated over and over again, the relative frequency of an outcome stabilizes toward the probability of that outcome, as illustrated in the graph showing that the probability of heads is half.

Probability is about determining models that describe the long-term predictable patterns from a random process. Consider the coin flip example. The probability of getting heads can be described as the proportion of heads seen in the long run. In other words, to determine the proportion of heads, we could continually flip a coin, count the number of heads after each flip, and calculate the total number of times the coin showed heads divided by the total number of coin flips completed, as seen in the previous table and graph. This process is called a **simulation**. In this case, we actually performed the random process of flipping a coin a large number of times. After 1000 repetitions (trials), a head occurred in 475 of the trials. Thus, based on these results, we would estimate the probability of a head occurring in any trial as 0.475. This would give the relative frequency of the number of heads and give us an empirical probability. (Note that sometimes when one is able to carry out the random process explicitly, this is referred to as an experiment. The term *simulation* is reserved for mimicking a random process that cannot actually be carried out. See, for example, the following blueberry pancakes investigation). Carrying out this coin-flipping process, we would notice that as the number of flips increases, if the coin is fair, the proportion of heads (the relative frequency of heads) stabilizes around half. Therefore, we say that the probability of getting heads when flipping a coin is half.

A **simulation** is a model of a real-world phenomenon that mimics the possible outcomes of a random process. If the actual random process can be performed physically, then it is often simply referred to as an experiment. For the purpose of this book, we will refer to all random processes that are carried out multiple times as simulations, regardless of whether the process can be physically done or whether the process has to be mimicked in some way.

Well-designed simulations are those that match well the real-world scenario, whereas substandard simulations are those that are not modeling the real-world process. For example, consider the real-world scenario of selecting a national committee of four mathematics teachers from a group of 20 NCTM members who have already volunteered their time. To have a diverse committee, we want to ensure that the northern, western, eastern, and southern regions of the United States are represented. We want to know the probability of selecting the committee and obtaining representation from each of the different regions of the United States. Our investigative question is, "What is the probability of selecting a committee with one representative from each region?" To simulate the committee selection, we would have to know the regions of the 20 NCTM members and then select four of those members at random. To simulate this selection process, we could use a deck of cards and represent each of the 20 NCTM members and their region with a separate card suit. For example, clubs could represent the South, heats could represent the East, diamonds the North, and Spades the West. Each card would represent a teacher. Then, we could mix the 20 cards and select four at random. This would model the random selection of a committee. If we repeatedly shuffle the cards and deal four cards, then we can simulate the long-term behavior of the random process of selecting four teachers at random from the original group of 20 teachers. The key in this simulation would be setting up the deck of 20 cards in such a way that one card would represent one of the 20 people (a substandard simulation might instead merely take any 20 cards and then select four, not recognizing that the four suits correspond to each of the four regions).

Becoming proficient in designing simulations is not an easy task. Students and teachers alike need extensive practice modeling real-world scenarios using different manipulatives and eventually software. The essence of probability modeling (or probabilistic modeling) is to determine a model that describes the long-run proportion of times an outcome should occur if the random process generating the outcome is repeated a large number of times. For example, when we think of tossing a fair die, we say that the probability of rolling a two is one-sixth. We are able to arrive at the mathematical model P(tossing a 2) = 1/6 because of the assumption that the die is fair. In the context of probabilistic modeling, this means that if we were to roll a die over and over again many times, we would expect, on average, to see that about one-sixth of the time, a two would be rolled. It does not tell us that if you roll a die six times, you will get a two one time within the six rolls.

Throughout this unit, we investigate the ideas of probability through the empirical lens. We will distinguish between the **theoretical probability** and **empirical probability** of an outcome, and show how they are related. **Theoretical probability** is the *actual* probability of an outcome (also often referred to as the **mathematical probability**). We demonstrate these ideas in all the investigations in this unit.

The following three units will make the connection between probability and statistics explicit. Namely, we show how to (1) use probability in statistics to model data from situations in real life, (2) use empirical probabilities to estimate actual probabilities and make statistical predictions about what is likely and what would be unusual, and (3) use probability to characterize the behavior of sampling distributions to help us draw inferences. Unit 3A focuses on (1) and introduces the ideas of (2). Sampling distributions (3) will be discussed in Units 3B and 3C.

Investigation 3A.1: Which Deck of Cards Is Fair?

Goals of this investigation: Define the probability of a random event as the long-run frequency, the law of large numbers, and experimental probabilities converging on mathematical probabilities.

A teacher in a class prepares two decks of 26 cards each. Deck one has 13 black cards and 13 red cards. deck two has 19 black cards and seven red cards. A class of 30 students is asked to determine which deck is fair, where fair is defined as having the same number of red cards as black cards. The students decide that each student in the class will draw 10 times with replacement from each deck and record their outcomes in a table. The catch is that they can look at only one randomly selected card at a time.

Deck 1	Deck 2

The investigative question posed to the class is:

Which deck is fair?

Tables from three different students are:

Student A		Student B		Student C	
Deck 1	Deck 2	Deck 1	Deck 2	Deck 1	Deck 2
R	B	R	R	B	R
R	B	B	B	B	B
B	B	B	B	R	B
B	B	B	R	R	R
R	B	B	B	R	R
B	B	B	B	R	B
B	R	B	B	B	B
B	B	B	B	R	B
B	R	B	B	R	R
B	B	B	B	R	B

Each student is asked to look at their outcomes and predict which one of the decks was the fair deck. The notion of "fair" in this case is interpreted as the chance of getting a black card being the same as the chance of getting a red card. However, as seen in Student A's table, in both cases red appeared fewer times than black (deck one had three red and deck two had two red draws). Does this indicate that both of the decks are not fair?

The random process is the act of drawing a card. Each draw has a random outcome: getting a black card or getting a red card. We denote that the probability of obtaining a red card corresponds to the relative frequency of getting a red card *in the long run* if the cards were replaced and shuffled in the deck after each draw. In other words, this probability can be expressed as a limit as the number of draws goes to infinity of the relative frequency. For teachers familiar with limit notation, the following expression illustrates the probability as a limit:

$$P(\textit{red card}) = \lim_{n \to \infty} \frac{\#\ \textit{of red cards drawn}}{n}, \text{ where } n \text{ is the number of draws.}$$

For n = 10, we can compute the relative frequencies of red cards obtained for each of the three students and each of the two decks as:

	Student A	Student B	Student C
Deck 1	3/10	4/10	6/10
Deck 2	2/10	4/10	3/10

To get a better sense of what happens to the relative frequency of red cards when the number of draws increases, the three students can combine all of their draws into one data set. This would provide the following relative frequencies:

	Pooled Data
Deck 1	13/30
Deck 2	9/30

Looking at these relative frequencies, we see that a red card was drawn from deck one approximately 43% of the time, and a red card was drawn 30% of the time from deck two. The relative frequencies fluctuate from student to student. The idea of probability is to see where these relative frequencies stabilize as the number of draws increases. In other words, the probability is the value of the relative frequency as the number of draws goes to infinity (as noted in the limit equation). Combining the data from all 30 students each drawing 10 cards from each of the two decks (300 total draws from each deck), the students obtain the following relative frequency of red cards:

	Pooled Data
Deck 1	153/300
Deck 2	81/300

From these pooled data, the students see that deck one is close to having red being drawn almost half of the time ($\frac{81}{300} = 0.51$). We can predict that as more and more draws are included in the data, the relative frequency of red draws for deck one will settle at half, which is considered the fair outcome, since each of the possible random outcomes (red or black) have an equal probability of occurring. In the case of deck two, we see that when more draws are included, the relative frequency of red stabilizes at around 0.27 ($\frac{81}{300}$).

If we assume that fair is 50–50, our results show that it is plausible that deck one is fair. Deck two has an empirical probability from our simulated data of 27%. This probability seems unusual if 50--50 is fair. Therefore, the class simulation indicates that it is not as plausible that deck two is a fair deck. For deck two, the likelihood of pulling a red versus a black card was not equally likely based on our simulated results.

While the pooled data table provides the relative frequencies (empirical probability), it does not represent the probability as previously defined. To determine probability, the students would have to draw infinitely many more times from each of the two decks.

> **INVESTIGATION SUMMARY:**
>
> **The main concepts developed in the fair coin investigation are:**
>
> 1. The probability of a random event is the long-run chance of that event happening after repeating a random process a large number of times.
> 2. Probability describes predictable patterns of random events.
> 3. Probability quantifies which events are plausible and unusual.
> 4. Random events can have outcomes that are equally likely or not.

The main point of this initial investigation is for students and teachers to understand the definition of probability and realize how we can use empirical probabilities to estimate and make statistical predictions about what is plausible and unusual. We want students and teachers to understand that in the short run, an outcome might not be predictable, but it may be in the long run. Drawing cards (playing cards or index cards), flipping coins, and tossing dice are traditional ways to approach teaching the definition of probability, because these manipulatives are typically easily available in classrooms. Statistics estimates the likelihood outcomes based on how plausible they are to have occurred given the data we see. The probability of an outcome is the long-run relative frequency of occurrence of that outcome. Through simulation, this interpretation provides statisticians with one method to make these estimations. The next several investigations solidify the idea of probability as the long-run relative frequency of outcomes from a random process and further develop the idea of using simulation to estimate probabilities.

Investigation 3A.2: Blueberry Pancakes

Goals of this investigation: Reinforce the idea of the long-run frequency as the definition of probability*, and simulate a random process using tools.*

The Blueberry Pancake House (BPH) prides itself on having at least one blueberry in all the pancakes they serve. In fact, their slogan is "No Pancake Should Be Without a Blueberry!" BPH wants to be able to advertise that its customers are likely to have blueberry pancakes that actually contain blueberries. BPH would like to give a predictive percentage of how often that should happen. The restaurant knows that the number of blueberries in a pancake is going to vary. The variability will fluctuate from batch to batch. Instead of cooking thousands of batches, BPH asks a sixth-grade class to model this situation and develop a percentage that the restaurant could use in its advertising. BPH tells the class that for an order of six pancakes, it uses 20 blueberries. The restaurant also mentions that it uses a robot arm to make the orders. The robot arm dips a specific-size ladle into a large bowl of batter to make each pancake.

The investigative question posed to the class is:

What is the probability of obtaining a pancake with no blueberries in an order of six pancakes?

To answer this question, the class is asked to brainstorm ideas of how to use materials that are available in the classroom to model the random process of the robot arm making six pancakes with 20 blueberries. One student has the following idea:

1. Distribute a sheet that has six circles drawn and labeled 1, 2, 3, 4, 5, and 6, representing the six pancakes.
2. Provide a die to each student in the class.

Assume the robot arm reaches into the batter bowl and fills its ladle with a randomly selected portion of the batter. To perform a simulation, the class must make basic assumptions about the random process. The random process the class will model corresponds to assigning each of the 20 blueberries to one of the six pancakes. This process can be modeled in the following way:

1. Consider the first blueberry. Roll the die, and whichever value comes up, place the blueberry in the circle that is labeled with that number.
2. Repeat this process for each blueberry (20 times).

In this manner, each student in the class will **simulate** the random process of assigning the 20 blueberries to pancakes as it would happen at BPH. For example, one student obtained the following picture, where each star represents a blueberry and how it was assigned:

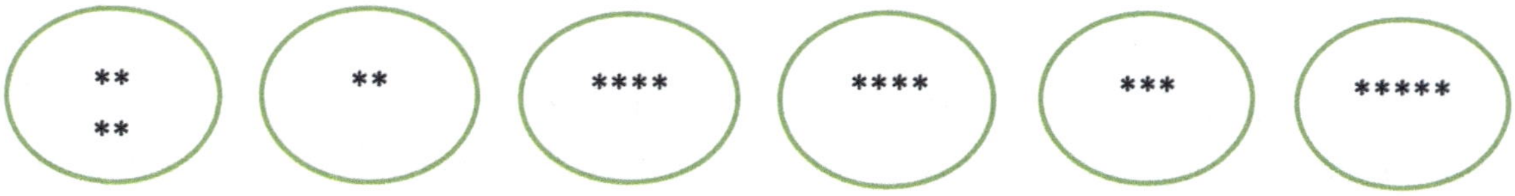

In this picture, all of the pancakes received at least two blueberries. For each simulated batch of pancakes, we can record the number of blueberries that pancake will have.

From this, we can determine whether or not every pancake has no blueberries or at least one blueberry. It is important to draw attention to the fact that the simulation entails assigning all 20 blueberries, not just one individual blueberry. The class is repeating the process of cooking an order of six pancakes with the 20 blueberries in them. For each simulated batch of six pancakes, the students can then record whether each pancake in fact received no blueberries or at least one blueberry in the batch. Thirty such simulation results are recorded in the following table.

Student	Whether or Not Simulation Resulted in a No-Blueberry Pancake	Student	Whether or Not Simulation Resulted in a No-Blueberry Pancake
1	All had at least one blueberry	16	No-blueberry pancake occurred
2	All had at least one blueberry	17	All had at least one blueberry
3	All had at least one blueberry	18	All had at least one blueberry
4	All had at least one blueberry	19	All had at least one blueberry
5	All had at least one blueberry	20	All had at least one blueberry
6	All had at least one blueberry	21	All had at least one blueberry
7	All had at least one blueberry	22	All had at least one blueberry
8	All had at least one blueberry	23	All had at least one blueberry
9	All had at least one blueberry	24	All had at least one blueberry
10	All had at least one blueberry	25	No-blueberry pancake occurred
11	All had at least one blueberry	26	All had at least one blueberry
12	All had at least one blueberry	27	All had at least one blueberry
13	All had at least one blueberry	28	All had at least one blueberry
14	All had at least one blueberry	29	All had at least one blueberry
15	All had at least one blueberry	30	All had at least one blueberry

From this, we see that the relative frequency of obtaining a no-blueberry pancake for this class was 2/30 = 0.067. In only two cases did the simulation lead to a student getting a pancake order with at least one pancake with no blueberries. On the other hand, the class data might reveal that getting at least one pancake with three of the 20 blueberries on it is very likely, or that having one blueberry on a pancake is less likely.

We refer to the .067 as the empirical probability of obtaining at least one pancake without any blueberries. It is not the theoretical probability; to find that, the robot arm would have to make the six pancakes with 20 blueberries infinitely many times. It is impossible to carry out the actual experiment using the robotic arm making pancakes; thus, it is necessary to use a simulation. The empirical probability offers a glimpse at the long-term behavior of the relative frequency of getting zero blueberries in a pancake. In the long run, the empirical probability will converge on the theoretical probability. This is the essence of the law of large numbers, which will be further discussed in the next investigation.

Based on the students' study, BPH estimates that more than 93% of pancake orders of six pancakes will have at least one blueberry on each pancake!

In this investigation, we used dice to simulate the blueberries being randomly assigned to the different pancakes. The random process of assigning blueberries to pancakes is modeled using the die. The probability of the event of no blueberries in a pancake is given as the long-run outcome of this process. The probability in this example could also have been computed mathematically. Statistics focuses on drawing conclusions in the presence of randomness, and probability shows which events are likely and which are unlikely[1].

INVESTIGATION SUMMARY:

The main concepts developed in the blueberry pancakes investigation are:

1. Probability shows which events are plausible and which events are unusual.
2. A simulation is an action that mimics the random process using physical tools (e.g., dice) or software. A simulation is a model of the random process.
3. When a probability is not known, a simulation can be used to help see the long-run patterns.
4. The empirical probability of an outcome of a random process is the relative frequency of that outcome for a fixed number of trials.
5. The theoretical probability of an outcome of a random process is the relative frequency of that outcome as the number of trials tends to infinity.
6. Probability can be used to model real-world scenarios.

The prior two investigations have introduced one interpretation of a *probability* as the long-run relative frequency of a random event. In addition, in both investigations, we used simple simulations (one using coins and one using dice) to help us estimate the probability. Our simulations modeled the random processes and helped us see the overall patterns. The next three investigations emphasize the usefulness of using simulations to model random processes, and they introduce statistical software that can be used to repeat the simulation many times. We provide three examples to illustrate a variety of activities one could do to solidify these concepts; however, all three of the following activities have similar goals, so if one were short on time, one could cover only one of these activities.

1 Note that an outcome is one individual result that can occur after performing the random process. An event consists of a collection of outcomes.

Investigation 3A.3: The Last Banana

Goals of this investigation: Use simulations to estimate probability and make decisions.

Two people are on a deserted island, and they decide to play a game to see who will get the last banana to eat. They have two dice with them on the island. The game is as follows:

Both players roll their die. If a one, two, three, or four is the highest number rolled, then player A wins. If instead a five or a six is the highest number rolled, then player B wins[2].

The investigative question is:

Which player would you rather be?

To begin the game, one can have students predict which player they want to be. Students will tend to choose player A because they believe there are more chances to win with four possible outcomes on one die that are winners (one, two, three, and four), versus only two possible outcomes that are winners for player B (five and six). Students might believe they have two-thirds probability of winning being player A and one-third probability of winning being player B.

2 Images taken as snapshots from the TED Ed talk presenting this problem. Retrieved here www.youtube.com/watch?v=Kgudt4PXs28.

Students can be paired together to play the game and asked to simulate the game 20 times. An example of tallies for 20 simulated games is:

Highest Value Rolled	Winner	Highest Value Rolled	Winner
4	A	6	B
4	A	6	B
6	B	5	B
5	B	5	B
6	B	2	A
6	B	1	A
5	B	6	B
4	A	1	A
1	A	1	A
1	A	6	B

In this example, the game was simulated 20 times. We see that A wins nine times out of the 20 and B wins 11 times out of the 20. Are students convinced by this evidence that they want to be player B?

To further explore, we can begin by combining all the data collected from each set of partners and computing the proportion of times A won in the class and the proportion of times B won in the class; with an advanced class, one can move immediately to using software and simulate the game 10,000 times. Pedagogically a teacher should gauge the class as to whether the additional step of combining the class's data is needed in this investigation or whether the students are conceptually ready to move to technology at this point. If a class needs more practice with the concepts, then completing the step of combining the class's data is suggested. If the class instead is comfortable with the ideas discussed and presented, then a teacher can move directly

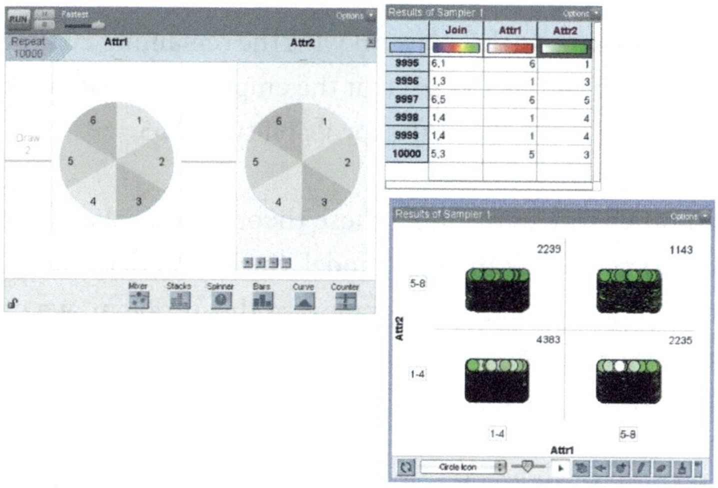

to using software for a large number of simulations. Using a statistical software, such as TinkerPlots, we can simulate the game 10,000 times . The previous screenshot displays possible results.

What we have done in this TinkerPlots window is set up two spinners with six possible outcomes that are equally likely to simulate the tossing of two dice. Then we run the spinners 10,000 times and we see that a five or six was rolled in die one but not die two a total of 2239 times. We got a five or six on die two and not die one a total of 2235 times, and we got a five or six on both dice 1143 times. Therefore, the empirical probability of player A winning in this simulation is $\frac{4,383}{10,000} = 0.44$ and the empirical probability of player B winning in this simulation is $\frac{2,239 + 1,143 + 2,235}{10,000} = 0.56$.

We can also approach this problem by computing the theoretical probabilities using formulas and probability rules. To do this, first we define the **sample space** for the random outcome of rolling two dice. The **sample space** is the set of all possible outcomes of a random outcome. In this case, one could get when the two dice are rolled. It is:

(1,1) (1,2) (1,3) (1,4) (1,5) (1,6)
(2,1) (2,2) (2,3) (2,4) (2,5) (2,6)
(3,1) (3,2) (3,3) (3,4) (3,5) (3,6)
(4,1) (4,2) (4,3) (4,4) (4,5) (4,6)
(5,1) (5,2) (5,3) (5,4) (5,5) (5,6)
(6,1) (6,2) (6,3) (6,4) (6,5) (6,6)

The red potential outcomes are those where player A would win, while the others are those where player B would win. We see that player A wins 16 out of the 36 possible outcomes. Therefore, P(player A wins) = 16/36 = 0.45.

We also observe that player B wins the remaining 20 times. Therefore, P(player B wins) = 20/36 = 0.55. We can see that the empirical probability computed from the simulation is very close to the theoretical probability of 0.55.

Another way to compute these theoretical probabilities is to recognize that the list of outcomes is just an array model. Such a model is often used when we teach multiplication in the elementary grades. In this problem, we can think of the red outcomes as representing when A wins and the other outcomes as representing when B wins. From this, we can see that:

P(A wins) = P(die 1 has an outcome of 1, 2, 3, or 4) * P(die 2 has an outcome of 1, 2, 3, or 4)

This is the multiplication formula for two **independent** events. We say that two events are independent if the outcome of one of the events does not affect the outcome of the other event. In the case of the dice rolls, the outcome of the first die does not affect the outcome of the second die in any way. An example of dependent events might be the event of picking a spade out of a deck of cards on a first pick and then picking another spade on a second pick. Because one card is already chosen from the deck, the probability of getting certain cards changes for the second pick. These events are dependent. More discussion of independence in the context of two-way tables is included in later units.

From the array, P(dice 1 has an outcome of 1, 2, 3, or 4) = 4/6 and P(dice 2 has an outcome of 1, 2, 3, or 4) = 4/6.

Therefore, $P(\text{A wins}) = \left(\frac{4}{6}\right) \cdot \left(\frac{4}{6}\right) = 0.45$.

Using the array representation, we notice that the P(A wins) + P(B wins) = 1, the entire space. We can then find P(B wins) = 1 - 0.45 = 0.55. This investigation thus offers a way to teach the multiplication formula in probability through simulation.

The theoretical probabilities computed closely match the empirical probabilities that were simulated after 10,000 simulations. At this point, the students should be convinced that they would want to be player B.

A video for this investigation can be found at:
https://ed.ted.com/lessons/the-last-banana-a-thought-experiment-in-probability-leonardo-barichello#review.

A series of nine follow-up questions are also listed at the link (questions six through nine are adapted as Follow-Up Question 1 below).

This activity has been developed by many. It is presented in a TED Ed talk retrievable at www.youtube.com/watch?v=Kgudt4PXs28. It was also presented by Doug Tyson at the California Mathematics Council Conference in the fall of 2017.

As an additional note to this investigation, teachers could explore in more depth the law of large numbers. This would not be covered in the school-level curriculum, but could be introduced and covered for teachers. This investigation demonstrates the law of large numbers at play. The law states that the relative frequency of a specific outcome of a random process tends toward the theoretical probability of that outcome as the number of repetitions tends to infinity. In this example, suppose we notate the relative frequency of the number of times A wins out of n trials of the game as $\frac{A_n}{n}$ where A_n is the number of times A wins and n is the total number of trials. Then, the law of large numbers states the following:

$$\lim_{n \to \infty} P\left(\left|\frac{A_n}{n} - p\right| > \varepsilon\right) = 0$$

where p denotes the theoretical probability of A winning (in this game, that is 0.45).

This is saying that the probability that the relative frequency $\frac{A_n}{n}$ is different from p is tending toward 0. What is important to note in the law of large numbers is that it connects the empirical probability of an outcome to the theoretical probability of an outcome.

INVESTIGATION SUMMARY:

The main concepts developed in the last banana investigation are:

1. The empirical probability computed through simulations will converge on the theoretical probability as the number of simulations performed increases.
2. Illustrate how to estimate probabilities for complex situations using simulations.
3. If X and Y are independent events, then P(X and Y) = P(X) • P(Y).
4. The complement rule is given by the following formula: P(X) = 1 - P(not X).

Investigation 3A.4: Game Board[3]

Goals of this investigation: Introduce the notion of probability in a complex situation through simulations.

The committee for a school fundraiser is organizing a game to raise money for the school. To play the game, parents purchase tickets. For each ticket they purchase, they get one turn at the game. The game the committee designs is the following:

A wooden peg board is constructed and a player releases one ball from the top of the board. The ball follows some pathway and ends in a bin at the bottom of the board.

3 This activity was adapted from an activity developed by Wendy Weber at Central College, Iowa.

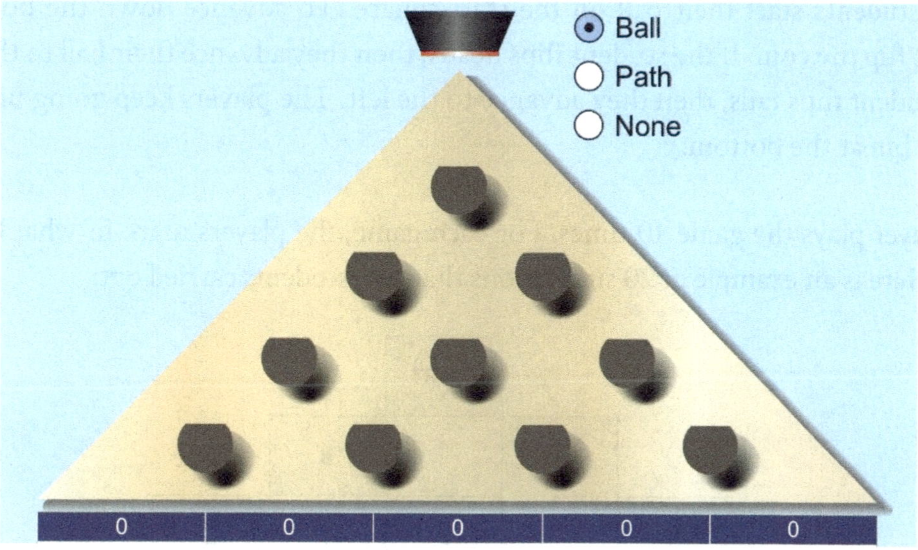

Image created via https://phet.colorado.edu/sims/html/plinko-probability/latest/plinko-probability_en.html.

The pegs on the board are placed in such a way that there is an equal chance for a ball to fall to the left past the peg or to the right past the peg. At the bottom, the cell has either a prize or no prize associated with it. A total of two prizes need to be placed at the bottom bins. The students of the school want to create a game so that the chances of winning a prize is low and those of not winning a prize are high. The students pose the following investigative question:

Where should the two prizes be placed in order to have the lowest probability of winning?

To help answer the investigative question, the students decide to simulate a few drops of the balls in the game board. They set up the following simulation:

Each student receives a diagram of the game board showing all possible pathways the ball could take, a coin, and a ball. The following picture shows the game board that is used by the students:

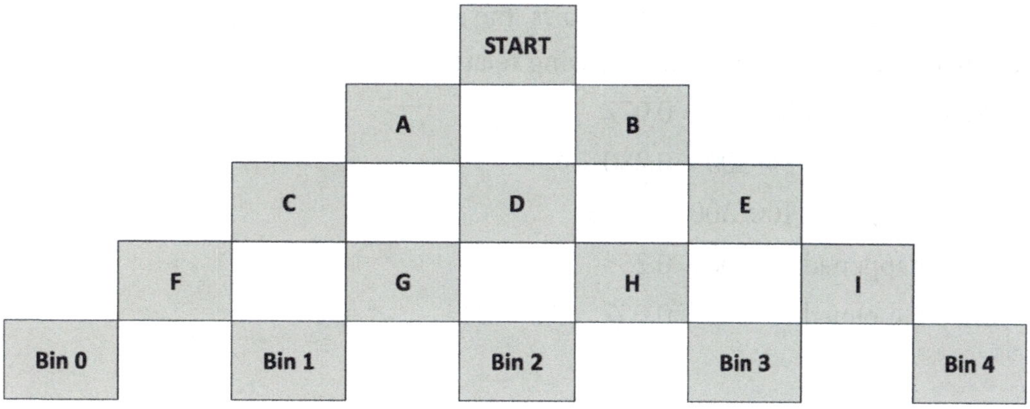

All the students start their ball on the start square. To advance down the board, the students flip the coin. If the student flips heads, then they advance their ball to the right. If the student flips tails, then they advance to the left. The players keep going until they end in a bin at the bottom.

Each player plays the game 10 times. For each game, the players mark in what bin they ended. Here is an example of 20 simulations that two students carried out:

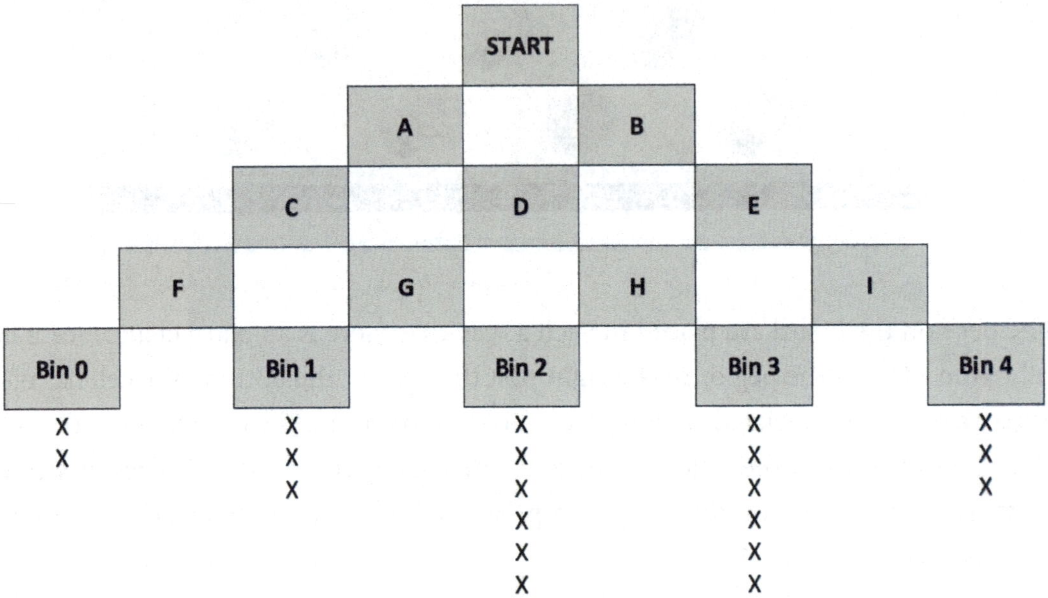

Based on the 20 simulated games, the students can compute the relative frequency of landing in each of the cells. For the simulated game depicted, the relative frequencies are:

- Bin 0 happened 2/20 = 0.10
- Bin 1 happened 3/20 = 0.15
- Bin 2 happened 6/20 = 0.30
- Bin 3 happened 6/20 = 0.30
- Bin 4 happened 3/20 = 0.15

After going through the simulation in pairs, the students decide to combine their results into a single data set. They get the following relative frequencies for the class:

- Bin 0 happened 23/300 = 0.077
- Bin 1 happened 84/300 = 0.280
- Bin 2 happened 100/300 = 0.333
- Bin 3 happened 76/300 = 0.253
- Bin 4 happened 17/300 = 0.057

While having 300 simulations is indeed a lot, the students want to find out if these relative frequencies hold in the long run. To test this, they find an online game that mimics their game at https://phet.colorado.edu/en/simulation/plinko-probability. They then run the simulation 10,000 times. The following results are obtained:

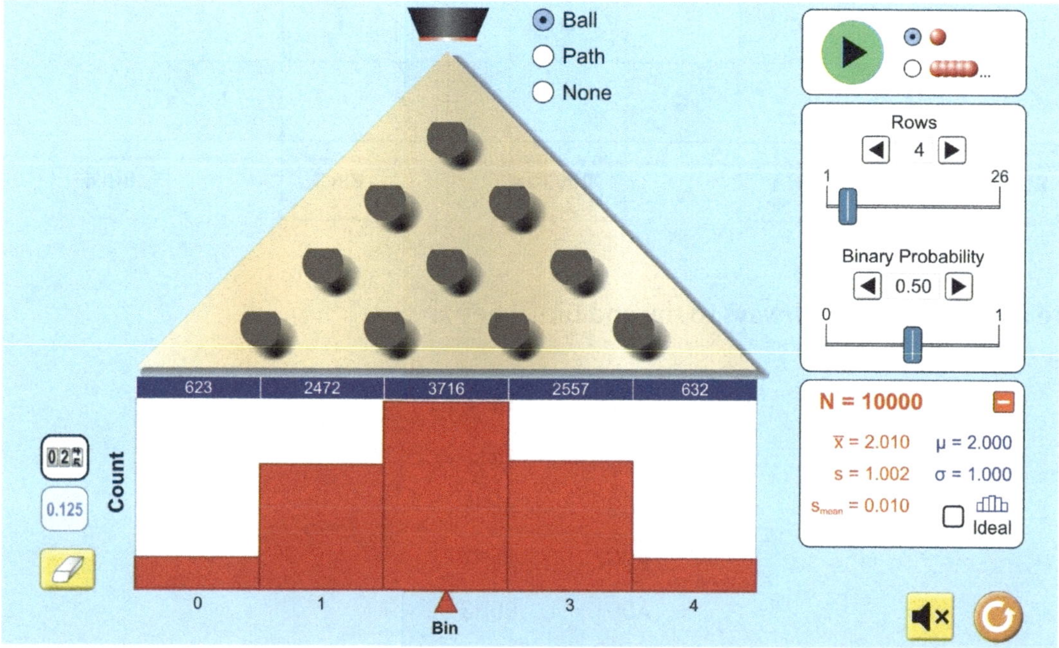

For this large simulation, the following relative frequencies and empirical probabilities are computed:

- P(Bin 0) = 623/10,000 = 0.062
- P(Bin 1) = 2472/10,000 = 0.247
- P(Bin 2) = 3716/10,000 = 0.372
- P(Bin 3) = 2557/10,000 = 0.256
- P(Bin 4) = 632/10,000 = 0.063

These represent the empirical probabilities of landing in each cell. Next, we can use formulas and rules to compute the theoretical probabilities. We then can compare how close our empirical probabilities are to the theoretical ones.

To begin, the students list their sample space, which consists of all possible pathways from the start to the bins at the end.

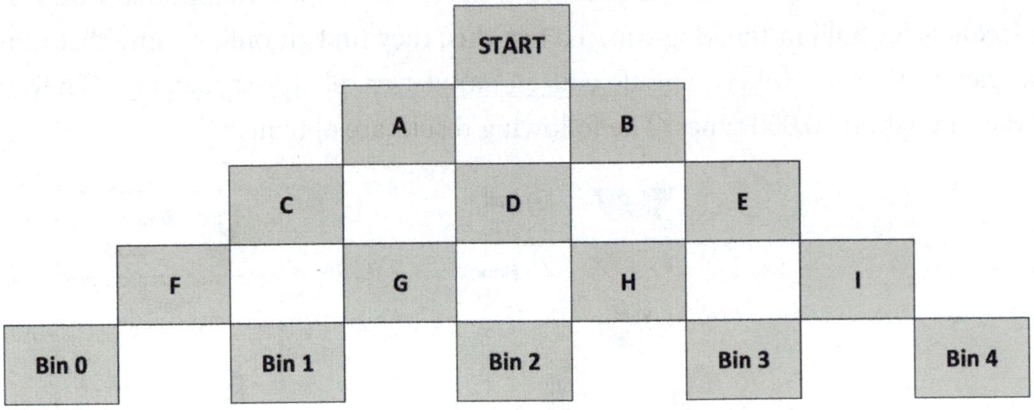

There are 16 possible pathways to the end bin. They are:

ACF0	BEI4
ACF1	BEI3
ACG1	BEH3
ACG2	BEH2
ADG1	BDH3
ADG2	BDH2
ADH2	BDG2
ADH3	BDG1

The first thing to notice is that although each pathway is equally likely (have the same probability), the probabilities of landing in different bins are not equally likely. Instead we see that many more pathways lead to bin two than bin zero. In fact, bins zero and four are the ending cell for one pathway, bins one and three are the ending cell for four pathways, and bin two is the ending cell for six pathways. From this, we can compute the following theoretical probabilities:

$$P(A) = P(E) = \frac{1}{16} + 0.0625$$
$$P(B) = P(D) = \frac{4}{16} = 0.25$$
$$P(C) = \frac{6}{16} = 0.375$$

We note that these theoretical probabilities were very well approximated by empirical probabilities found by the 10,000 simulations. From these results, the students decide to put prizes on bin zero and bin four only.

As an additional note to this investigation, teachers could explore the connections to the binomial coefficient and Pascal's triangle by considering a generalization of the Plinko board to incorporate more rows. The online simulation app allows one to increase or decrease the number of rows in the Plinko board. The board in the investigation has four rows with five bins; however, what would the probability of landing in specific bins be if there were seven rows? Ten rows? Fifty rows? rows? We can generalize our results to the case of rows using counting rules and noticing patterns.

First, if there are n rows in a Plinko board, how many bins will be at the bottom? The answer is that there are $n + 1$ bins.

Next, in the previous, four-row example, we found that there were 16 total possible pathways to get to the bottom. We found this by writing out all of the pathways; however, we could recognize that for four rows, we have a total of $2^4 = 16$ possible pathways. In the general case of n rows, this means that there are a total of 2^n possible pathways. This will give us our denominator for the theoretical probabilities.

For the numerators, we have to somehow express the number of pathways that lead to each bin in terms of n and the selected bin. As before, we can label our $n + 1$ bins from 0 to $n + 1$. By simulating, we notice that the middle bins always have more balls landing in them and that the number of pathways leading to the bins are symmetric about the middle bin(s). From this, we know that the number of paths leading to bin zero will be the same as the number of paths leading to bin $n + 1$, bin one is the same as bin n, bin three is the same as $n - 1$, etc.

We can first reduce the number of rows and look at the number of rows leading to the bins for these lower numbers. For example, for the Plinko board with two rows, we would have

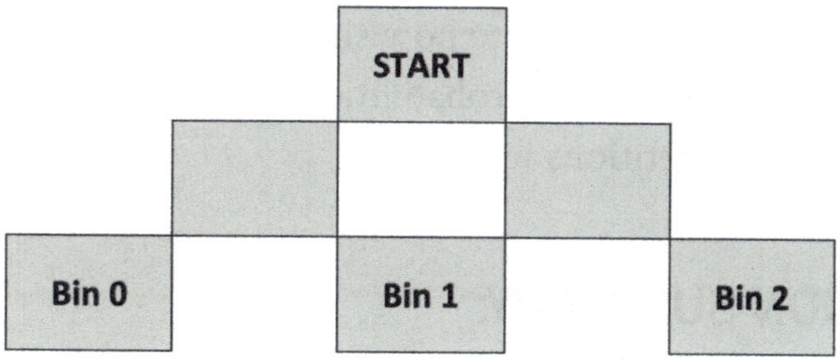

with one possible pathway leading to bin zero, two possible pathways leading to bin one, and one possible pathway leading to bin two. Adding one more row to the Plinko board, we have the following:

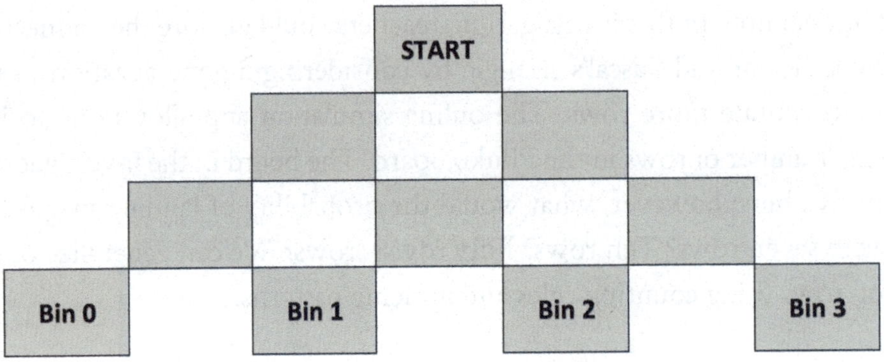

For this Plinko board, we can count the pathways again and we see that there is one pathway to bin zero, three pathways to bin one, three pathways to bin two, and one pathway to bin three.

At this point, we observe a familiar pattern begin to emerge—Pascal's triangle is at play! In fact, our example Plinko board with four rows has the possible pathways as one to bin zero, four to bin one, six to bin two, four to bin four, and one to bin four, which is exactly the next row in Pascal's triangle. Using our knowledge about Pascal's triangle, we know that the number of pathways to the $n + 1$ bins for the n row Plinko board must therefore be given by the binomial coefficient. Thus for bin k, the number of pathways leading to it in an n row Plinko board are:

$$C(n, k) = {}_nC_k \, \frac{n}{k} = \frac{n!}{k!(n-k)!}$$

These connections to the binomial coefficient and Pascal's triangle are additional mathematical ideas that can be easily connected to probability. These types of connections are not statistical in nature; however, they do offer math teachers and math-teacher educators ways to make deeper connections between mathematics and probability. While these concepts are interesting to explore, the main point of the game board investigation is not to derive or work with the binomial formula or Pascal's triangle, but instead to illustrate how simulations can be useful in computing probabilities and how the empirical probabilities tend toward the theoretical probabilities as the number of repetitions increases.

INVESTIGATION SUMMARY:

The main concepts developed in the game board investigation are:

1. Probability calculations can help make decisions.
2. Implementing hand simulations is useful to build conceptual understanding before moving to software to carry out a large number of simulations.

Investigation 3A.5: Random Exams

Goals of this investigation: Use probabilistic modeling and probability long-run computations through theory and simulation.

Jessica is a high-school teacher who recently collected and graded tests for her statistics class. While doing this, she noticed that out of all the tests she collected, four of them were turned in without names. When Jessica handed back the tests to the class, there were four students who did not receive a test, because they had forgotten to put their names on them. At this point, Jessica decides to randomly hand back the four tests to the four students and hope that the students get the test they turned in. She wonders what the chances are that each student will receive the correct test and decides to investigate the following questions with her class:

How often will all four students be given their correct test?

This same question could be rephrased as follows:

What is the probability that all four students received the correct test?

This scenario highlights probabilistic modeling. While the investigative questions ask for computations of probabilities, the underlying idea in the scenario is randomness. Students are being handed back tests randomly—a random process is occurring in this setting. The outcome from this random process of handing back the tests is not predictable, because we do not know whether each student will be handed back the correct test. The computation of the probability of obtaining four matches is based on the notion that if the random phenomenon of handing the tests back was repeated over and over again, we would want to identify the chances of getting four out of four correct matches.

With the class, Jessica can simulate the random process of handing back the tests over and over again and observe how many matches happen after each simulation, or she can compute the probability directly of four matches using theoretical rules. She begins by simulating the process over and over again, first by hand and then using software. She then can compare how close the empirical probability that she derived through the simulation matches up with the theoretical probability she computed using probability rules and formulas.

To simulate by hand, she takes four cards and labels them 1 through 4, each representing one of the tests, and four stick figures, labeled 1 through 4, each representing one of the four students (see Figure 1). She can mix the cards without looking at the labels and then deal the cards out to the labeled people. If the number on the card matches the person's label, then we call it a match. In the figure, we can see that the number of matches equals one, because person number 2 was the only individual to receive the correctly matched card labeled 2.

Figure 1

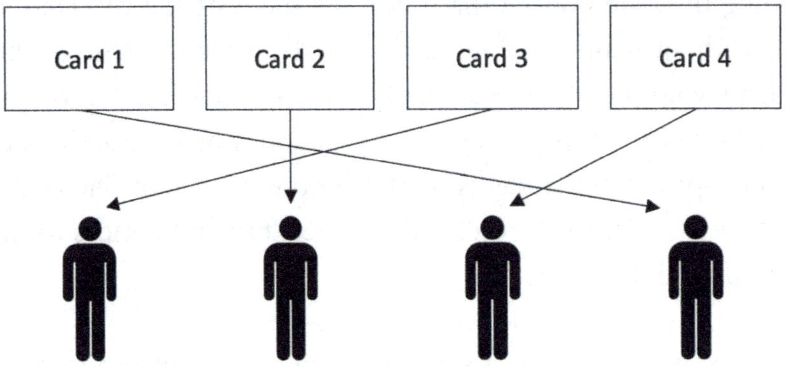

Next, she repeats this random dealing process 25 times. For each time, we could compute the number of matches made and then tally the total number of times that zero matches were made, one match was made, two matches were made, three were made, or four were made. To answer our investigative question, we would find how many matches of four were made and divide that number by 100. In this investigation, we move directly to software. For example, *R* can be used to carry out our simulation.

```
permutations <- permutations (n = 4, r = 4, v = 1:4, repeats.allowed = F)
```

First, Jessica simulates one round of passing back tests:

```
sample_permutations <- sample(permutations, size = 1, replace = TRUE)
```

Then, she wants to simulate 100 rounds of passing back tests. To do this, she first thinks about the sample space in order to set up the simulation correctly for the software. To consider all possible combinations of how the tests could be handed out, a sample space is notated. The **sample space** is the list of all possible outcomes of the random process. For this scenario, because there are four people, there are four "slots" to fill for each test. Suppose the teacher hands the first person test one, the second person test two, the third person test three, and the fourth person test four. This would give the outcome of 1234 and four people getting the correct four tests. Another outcome, 1342, would give the first person test one, the second person test three, the third person test four, and the fourth person test two. In this outcome, only person one received the correct test. After making a sample space with the 24 possible outcomes, a table, like the one shown previously, can be made to organize each outcome and the number of correctly matched tests.

Outcome	Number of Matches	Outcome	Number of Matches
1234	4	3124	1
1243	2	3142	0
1324	2	3214	2
1342	1	3241	1
1423	1	3412	0
1432	2	3421	0
2134	1	4123	0
2143	0	4132	1
2314	1	4213	0
2341	0	4231	2
2413	0	4312	0
2431	1	4321	0

After analyzing the table, the outcomes illustrate that all four tests being matched correctly (outcome 1234) occurred one time. The outcomes where two tests were matched correctly (outcomes 1243, 1324, 1432, 2134, 4231) occurred five times; one test matched correctly (outcomes 1342, 1423, 2314, 2431, 3124, 3214, 3241, 4132, 4213) occurred nine times; and zero tests matched correctly (outcomes 2143, 2341, 2413, 3142, 3412, 3421, 4123, 4312, 4321) nine times. It is worth noting that it is impossible to have three people obtain the correct test, but not the fourth. Therefore, the probability of three test-to-student matches occurred zero times.

We can simulate this process numerous times to see how many times the outcome 1234 occurs. This simulation can be done using code, software, or an online shuffling generator (e.g., www.dcode.fr/permutations-generator). For example, to run the simulation 25 times in R, she can use the following:

```
sample_permutations <- sample(permutations, size = 25, replace = TRUE)
```

She can also run the simulation 10,000 times using *R* and obtain the following results:

```
allcorrect <- sample_permutations[, 1] == 1 & sample_permutations[, 2] == 2 &
 sample_permutations[, 3] == 3 & sample_permutations[, 4] == 4
table(allcorrect)
FALSE  TRUE
 9604    396
```

The *R* code output shows that 396 times out of 10,000, the permutation 1234 randomly occurred. Carrying out a simulation can illustrate the probability of outcomes in an intuitive way. We refer to the probability 396/10000 derived by the simulation as the empirical probability of the outcome 1234. Here we showed the simulation coded in *R*, however, teachers and students not familiar with coding can do the simulation using other software or online tools. This code is shown as an example of how this simulation can be carried out.

An alternative to carrying out a simulation using software would be to compute the probabilities using theory[4]. For this scenario, Jessica can list out all the possible dealings that the four people could receive as outlined in the table. For example, person one could receive either test one, two, three, or four in one dealing. Then, depending on which test person one receives, person two can receive one of the remaining three tests. Person three can then receive one of the remaining two tests, and finally person four gets the last test. This means that there are $4 \times 3 \times 2 \times 1 = 24$ possible test dealings (four options for the first test, three options for the second test, two for the third test, and the remaining one for the fourth test).

Of these 24 possible test dealings, only one of them will have all four people receiving the correct test, namely the one where test one is dealt to person one, test two is dealt to person two, test three is dealt to person three, and test four is dealt to person four. Thus, the probability of obtaining four correct test-to-student matches is 1/24.

By counting up the number of times no matches occur, we calculate the probability of getting zero matches as 9/24. Similarly, by counting up the number of times one match occurs, the probability of getting one match is 8/24. And the probability of getting two matches is 6/24.

---

4 This same problem can be formulated using random variables, which are beyond the scope of this book. One can define a random variable X as the number of correct exams that are handed out. A random variable is a function from the sample space of the possible 24 outcomes to the set of numbers {0, 1, 2, 3, 4}. The random variable counts the number of correct matches. Then, the probability can be formulated in the following manner: P(four correct test-to-student matches) = $P(X = 4) = 1/24$. As noted, random variables are beyond the scope of this book. For readers with an interest in learning more about the theoretical formulation of random variables, other books can be consulted.

More formally, we can write the following:

$$P(\text{four correct test-to-student matches}) = \frac{1}{24}.$$

We observe that this theoretical probability in fact closely matches the empirical probability when we performed the simulation 10,000 times. While in the short run, with 100 trials being simulated by hand, gave a probability of 4/100, this probability settled to 1/24 as we increased the number of trials to 10,000. When we increased the number of times we simulated the random process, the empirical probability converged on the theoretical probability we found using probability formulas.

The class recommends that teachers not hand back exams randomly and that as students, they must always be conscious of writing their name on their test.

Allan Rossman and Beth Chance provide an applet for a similar scenario with random babies in a hospital not being labeled and then assigned to their mothers. The applet can be found here: www.rossmanchance.com/applets/randomBabies/RandomBabies.html.

The applet provides a way to simulate many trials and see the results.

INVESTIGATION SUMMARY:

The main concepts developed in the random tests investigation are:

1. A random process is a process for which the individual outcome is unpredictable. However, if the process is repeated a large number of times, then predictable patterns emerge in the outcomes.
2. Probability describes predictable patterns, and statistics uses these predicted patterns by comparing what is actually seen in data with what should be expected.
3. Probability modeling provides a model that describes the long-term outcomes from a random process.

The previous investigations illustrated several important ideas. The first key concept is that probability is associated with a random process or random phenomenon. In statistics, we talk about probability in the context of understanding potential outcomes to a random phenomenon. This random event can have many possible outcomes that we can simulate (or collect data on) to observe their occurrence. These collected data show the connection to statistics. Statistics requires us to examine data in order to understand how likely the data are to occur, which, in turn, can help us understand how to analyze and interpret the data.

Second, these investigations again highlight the idea of probability as a long-run relative frequency. The examples illustrate how in the long run, the relative frequency of an event "settles" to the theoretical probability of an event. This is a fundamental idea in statistics whereby we are trying to estimate things given the data being examined. One can compare the expected long-term patterns one should see in data with what one actually sees in the collected data in order to draw statistical conclusions.

Next, we introduce two commonly studied probability rules (the addition formula and the conditional formula) through simulations in order to show the usefulness of simulations to model more complex real-life situations, as well as to show how to make decisions and statistical predictions about what is unusual and plausible.

Investigation 3A.6: Soccer-Practice Game

Goals of this investigation: Introduce the addition rule through simulations.

A new high-school soccer team has been formed. To motivate the players, the coach plans to play a fun game at the end of each practice. The game consists of trying to score a goal from a corner kick and dribbling the ball in the air. To win the game each practice, a player needs to do one of the following:

- Make a shot directly on goal from a corner kick
- Dribble the ball in the air at least six times

At each practice, the team plays the game. There are a total of 100 practices throughout the season.

Before playing the game at a practice, Bella practices at a field by her house. She makes two shots directly on goal from a corner kick out of 10 attempted shots. She also dribbles the ball in the air six times without dropping the ball seven times out of 10 attempts. She then goes home and sets up a simulation of the 100 upcoming practices based on this preliminary practice to compute her probability of winning the game. She asks the following investigative question:

What is the probability of Bella winning the game?

She uses technology to simulate her performance in the game at each practice. She simulates the two components of the game independently. She makes an important assumption that corner-kicking and dribbling skills are not related; thus, modeling the two tasks as independent in the simulation. First, she sets up a spinner to represent her chances of making a corner kick. Based on her practice, she assumes that she has a 20% chance of scoring a goal directly from a corner kick. Second, she sets up a spinner to represent her chance of dribbling the ball in the air six times successfully. She does this with a 70% chance of success. She could have constructed spinners by hand and spun the spinners 100 times to represent each game she will have to play in each practice. The simulation yields the following two-way table:

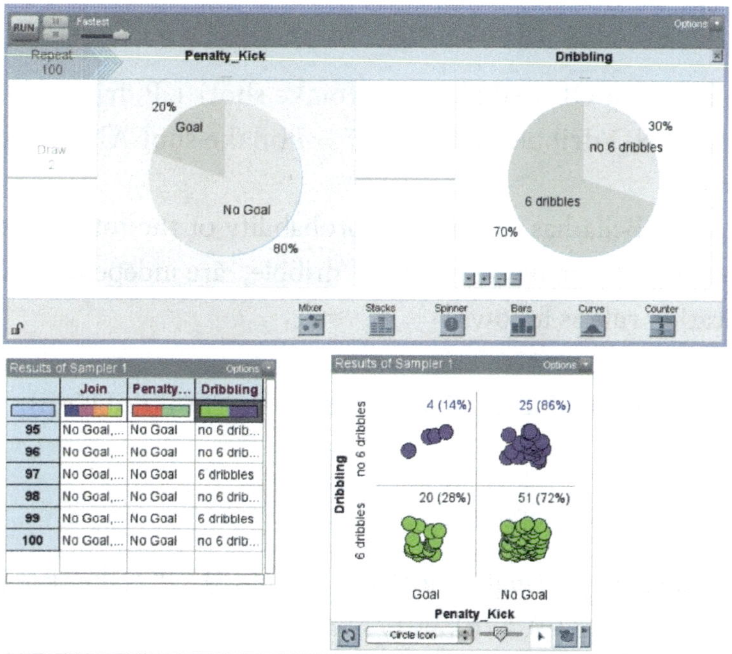

From the two-way table, Bella can compute *P(make shot OR dribble 6)*. She is successful at the game in three different scenarios: making the corner shot and not dribbling successfully, dribbling successfully and not making the corner shot, and making the corner shot and also dribbling successfully. The two-way table represents these scenarios in the top left corner (4), bottom left corner (20), and the bottom right corner (51). Therefore the probability can be found as follows:

$$\text{P(make shot OR dribble 6)} = \frac{20 + 4 + 51}{100} = \frac{75}{100}$$

Another way to find this probability from the two-way table is to note that:

$$P(\text{make shot OR dribble 6}) = P(\text{make shot}) + P(\text{dribble 6}) - P(\text{make goal AND dribble 6})$$

$$= \frac{24}{100} + \frac{71}{100} - \frac{20}{100}$$

$$= \frac{77}{100}$$

Bella could have also applied the mathematical addition rule instead of using a simulation to compute her probability. In applying the addition rule, Bella would have available only the estimated probabilities from her preliminary practice.

The addition rule yields:

$$P(\text{make shot OR dribble 6}) = P(\text{make shot}) + P(\text{dribble 6})$$
$$- P(\text{make shot AND dribble}) = 0.2 + 0.7 = P(\text{make shot AND dribble}) = ?$$

Given the information Bella has, to find the probability of the intersection, , she must assume that the two events, "make a shot" and "'dribble," are independent so that she can apply the multiplication rule as follows:

$$P(\text{make shot AND dribble}) = P(\text{make shot}) \cdot P(\text{dribble 6})$$

This then yields,

$$P(\text{make shot OR dribble 6}) = P(\text{make shot}) + P(\text{dribble 6}) - P(\text{make shot AND dribble})$$
$$= 0.2 + 0.7 - (0.2)(0.7) = 0.9 - 0.14 = 0.76$$

As we can see, the empirical probability found in the 100 simulation is very close to that found using the probability rule. This probability rule is called the **addition rule,** and it states the following: For two events X and Y, $P(X \cup Y) = P(X) + P(Y) - P(X \cap Y)$. If the two events are independent, then as noted in the prior investigations, $P(X \cap Y) = P(X) \cdot P(Y)$. Note that if the two events were dependent, then the simulation would be modeled differently. To create such a model, prior information would need to be known about the dependency.

INVESTIGATION SUMMARY:

The main concepts developed in the soccer-practice game investigation are:

1. The addition rule for events X and Y in probability states: $P(X \cup Y) = P(X) + P(Y) - P(X \cap Y)$.
2. If X and Y are independent, then $P(X \cup Y) = P(X) \cdot P(Y)$.
3. Simulations are useful to model and understand complex probabilistic scenarios.

Investigation 3A.7: Detecting Disease

Goals of this investigation: Introduce conditional probability through simulations.

Strep throat is a common virus that is often prevalent in schools. To detect whether someone has this virus, doctors administer a rapid strep test. Even though the rapid strep test helps doctors make a diagnosis, it does, at times, give the wrong results. Specifically, the test could come back negative when in fact a patient has strep (false negative) or the test could come back positive when in fact a patient does not have strep (false positive). As a doctor, it is important to understand the chances of these false positives and false negatives occurring so that risks can be communicated accurately with patients.

A doctor asks:

What is the probability of not having strep throat given that you get a positive rapid strep test result?

The doctor knows a few pieces of information to help her compute this probability (see www.medicinenet.com/rapid_strep_test/article.htm for reference):

1. Out of every 100 people that go to the doctor and get a rapid strep test, 75 people actually have strep throat then confirmed by a throat culture.
2. The rapid strep test has a **sensitivity** of 95%.
3. The rapid strep test has a **specificity** of 98%.

The **sensitivity** of a test refers to the ability of the test to correctly identify the patients who have strep. This means that the rapid strep test will be positive in 95 out of 100 patients who have strep throat. Five out of 100 patients with strep will be missed by the test, referred to as a "false negative."

The **specificity** of a test refers to the ability of the test to correctly identify the patients who do not have strep. This means that the rapid strep test will be negative for 98 out of 100 patients who do not have strep throat. Two out of 100 patients without strep will be false positives.

To investigate, the doctor starts by setting up a two-step simulation. She can model this situation with a hands-on simulation using different colored balls in bags, or she can directly go to technology to help her simulate a large number of trials. We will briefly describe a hands-on simulation one could go through, and then describe in detail a simulation that could be carried out using technology.

To simulate this scenario, the doctor needs 300 total balls and three bags. The first bag will model whether or not a person has strep throat, so she will include in the bag 100 total balls, of which 25 are green and 75 are red. The red ball symbolizes the people with strep throat, and the green balls those without strep throat.

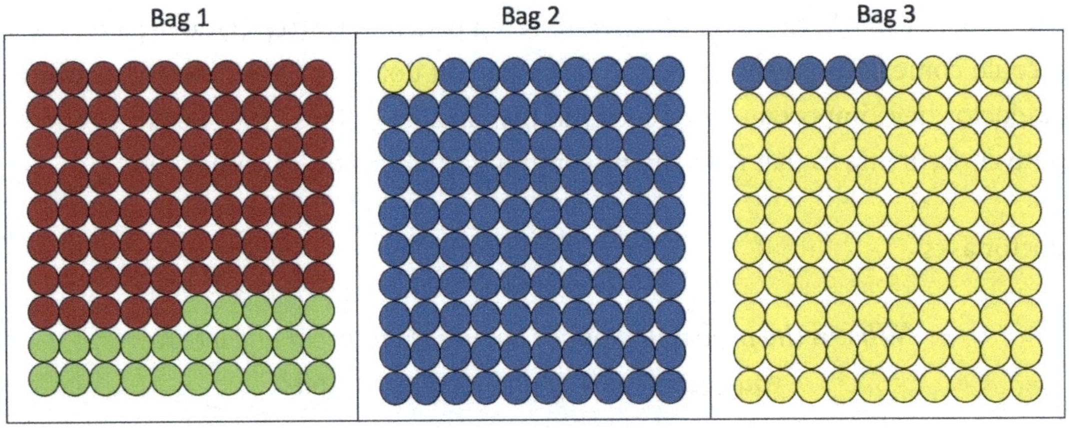

In the next bag, she also places 100 balls. Of these 100, she places two yellow balls and 98 blue balls. This bag represents the potential outcomes of a test when a person does *not* have the disease. The blue balls represent the test coming out negative, and the yellow balls represent the test coming out positive. In the last bag, she places the remaining 100 balls. This bag represents the potential outcomes of a test when a person *has* strep throat. In this case, there is a 95% chance the test will be positive, so she places 95 blue balls, and there is a 5% chance the test comes out negative, so she places five yellow balls.

With this setup, students can begin to do a two-step simulation by first picking out of bag one and then, depending on the outcome, picking out of bag two or three. For example, a student may pick

Bag one pick: Green ⟶ Bag three pick: yellow

or

Bag one pick: Red ⟶ Bag two pick: yellow

Students can simulate this process by hand by choosing the balls, and then placing them back in the bags and reshuffling. The following table illustrates how the frequencies for each of the possible outcomes would be recorded in a two-way table of this type:

	Negative Test	Positive Test
No Disease		
Disease		

There are four possible categorical outcomes (no disease and negative test; no disease and positive test; disease and negative test; and disease and positive test).

If instead we wish to move straight to software, we can set up a simulation using, for example, TinkerPlots that mimics the one we designed with the balls. Here is an image of three spinners, each spinner representing one of the bags:

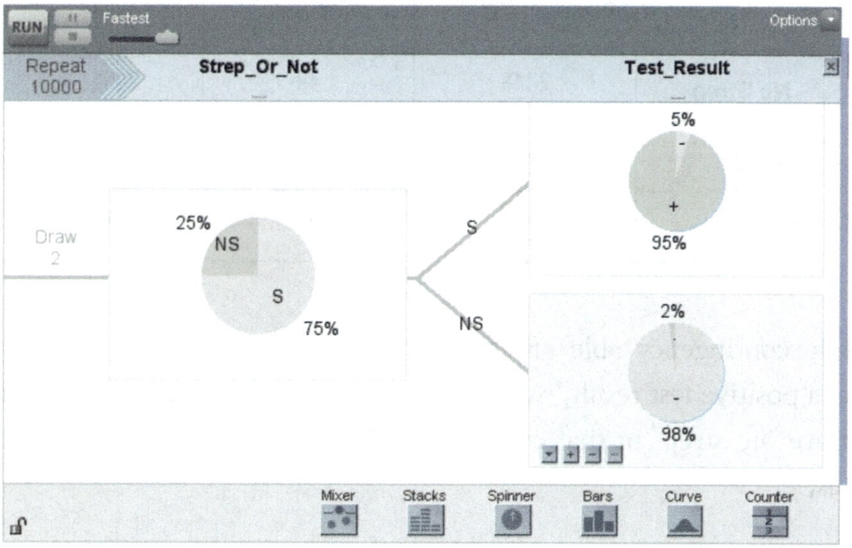

In TinkerPlots, we can run the simulation 10,000 times in a matter of seconds to obtain the following image:

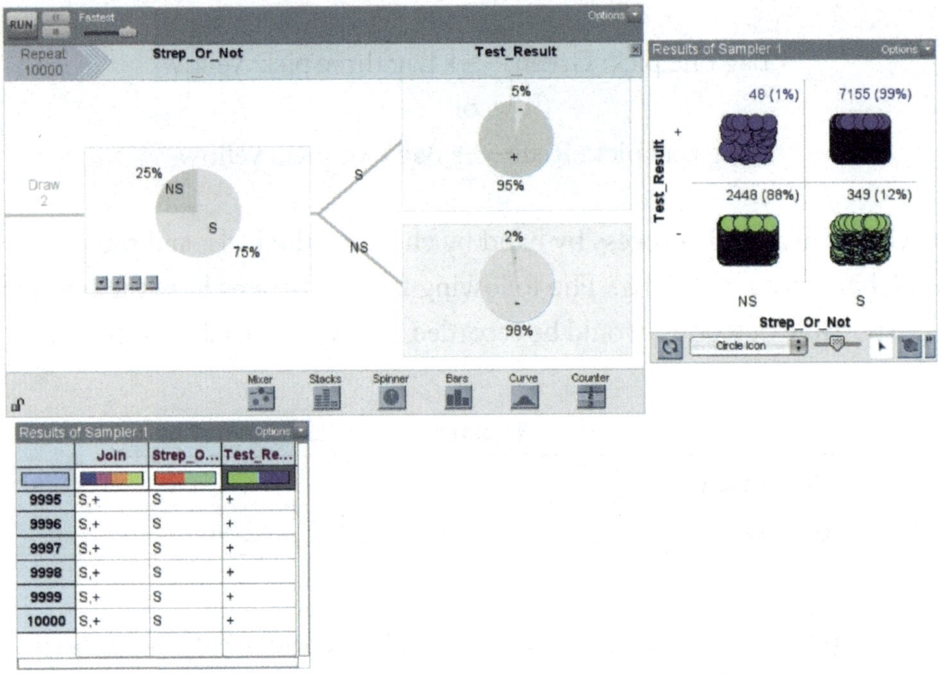

The table sorts the outcomes into the four possible outcomes that could occur. We can then transfer these outcomes into a contingency table similar to the one outlined previously. The table reports the frequency of the four categorical outcomes.

	Negative Test	Positive Test	Total
No Strep	2448	48	2496
Strep	349	7155	7504
Total	2797	7203	10,000

Looking at the contingency table (also referred to as a two-way table), when we say "given that you get a positive test result," we are contingent on the positive column in the table. Those that are "no strep" in that column are the 48 out of the total 7203 positive test results. Thus,

$$P(\text{no disease given } +) = P(\text{no disease} \mid +) = \frac{48}{7203} = 0.0066$$

We can also model this probability using a tree diagram to see all the possible outcomes and the probability of those outcomes:

Figure 2. Test Outcomes Tree Diagram

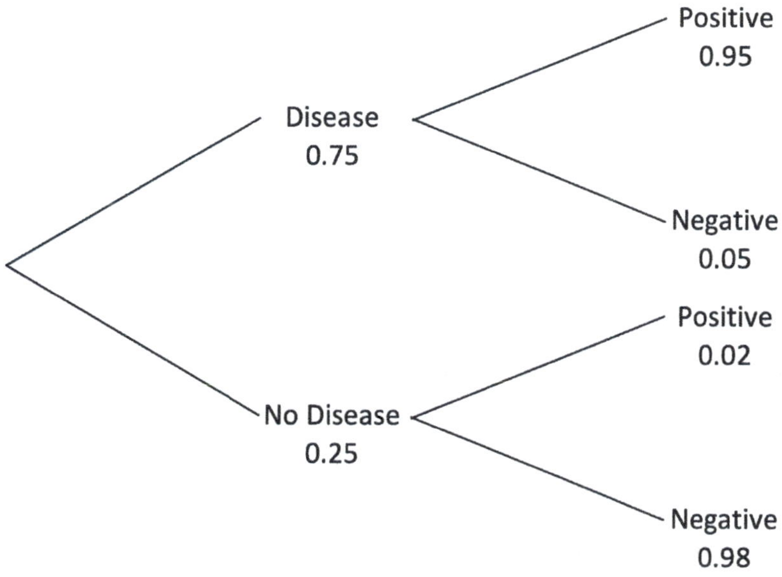

From Figure 2, we see that we have four possible outcomes:

- Disease and + test
- Disease and – test
- No disease and + test
- No disease and – test

The first pathway occurs with probability: P(strep AND +) = P(disease ∩ +) = (0.75)(0.95).

The second pathway occurs with probability: P(strep AND -) = P(disease ∩ –)=(0.75)(0.05).

The third pathways occurs with probability: P(no strep AND +) = P (no disease ∩ +) = (0.25)(0.02).

The fourth pathway occurs with probability: P(no strep AND –) = P(no disease ∩ –) = (0.25)(0.98).

The probability can also be computed through the conditional probability formula, which states the following for two events X and Y: $P(X \mid Y) = \frac{P(X \cap Y)}{P(Y)}$

For our problem, the formula yields the following results:

$$P\,(\text{no strep} \mid +) = \frac{P\,(\text{no strep} \cap +)}{P(+)} = \frac{P\,(\text{no disease} \cap +)}{P(+)}$$

We use our probabilities in the conditional probability formula in the following manner:

$$P(\text{no strep} \mid +) = \frac{P(\text{no strep} \cap +)}{P(+)} = \frac{P(\text{no disease} \cap +)}{P(+)} = \frac{(0.25)(002)}{(0.25)(0.02) + (0.75)(0.95)} = 0.0069$$

This gives us the probability of not having strep throat given that the test result came out positive as 0.69%. Note that when using our conditional probability formula, we do not need to simulate the outcomes. The conditional probability formula gives us the theoretical probability, and the probability computed through the simulation is our empirical probability. Notice again that these probabilities are very similar; in fact, they are almost identical. (For the simulation, the probability is 0.0066, and the theoretical computation leads us to a probability of 0.0069.) We see that the empirical probability converges on the theoretical probability.

In these more complex situations, simulations can help us "see" the process and convince us of the probabilities. While in simple probability examples our intuition helps us, oftentimes our intuition in slightly more complex scenarios is not accurate. Through simulation, however, we can convince ourselves and students of the computations. This provides rich activities for students instead of merely expecting them to memorize formulas.

INVESTIGATION SUMMARY:

The main concepts developed in the detecting disease investigation are:

1. The conditional probability of X given Y is given by: $P(X \mid Y) = \frac{P(X \cap Y)}{P(Y)}$
2. Simulations in complex, sometimes counterintuitive situations can greatly help clarify and convince us of the probabilities.
3. Using simulations as a model of a scenario is useful to understand complex probabilistic outcomes.
4. Using the visualizations of two way tables and tree diagrams promotes better conceptual understanding of conditional probability than using formulas.

Follow-Up Questions

1. Suppose person A and person B are on a deserted island, and there is one banana remaining. They each have a die and have to play a game to determine

who gets to eat the last banana. The game goes as follows: Each player rolls a die and the outcomes are multiplied. If the product is a multiple of six, player A wins. If the product is not a multiple of six, then player B wins. Which player would you rather be?

2. Design a simulation for the scenario. Carry out the simulation for 20 iterations by hand. Compute the theoretical probabilities associated with the scenario and discuss the relationship between the empirical and theoretical probabilities you found.

   A professional society of teachers wants to select a two-person subcommittee for its curriculum committee. There are eight candidates, four from the southern region and four from the northern region, whom it feels would serve well on the subcommittee. If the committee would like to select one member from each of the northern and southern regions to ensure regional representation, what are the chances that the two-person subcommittee will ensure regional representation? This same question could be rephrased as follows:

   What is the probability that the two-person subcommittee will have one member from the southern region and one member from the northern region?

The investigations in this unit offer a deep and thorough introduction to probability through the use of simulations. The investigations in this unit intended to demonstrate the connection between probability and statistics by using probability to model data from situations in real life, and using empirical probabilities to estimate and make statistical predictions about what is plausible and unusual. Again, statistics is a way to draw conclusions in light of randomness, and probability gives you the tools to model and quantify randomness.

We notice that often our intuitions in complex probabilistic scenarios are not correct. Simulations offer us a way to explore the outcomes of complex situations and check our intuition. It is important to realize that using simulations through technology as described in this unit is reliant on the capabilities of software.

When probabilities were computed before software was available, it was often not feasible to simulate a random process many times (e.g., 10,000 times). Thus, more reliance was placed on the formulas and theoretical rules than they are today. The beauty of simulation is that it makes probabilistic modeling accessible to students in K–12, not just postsecondary students.

Simulations are explicitly mentioned in state standards in middle school, and probability rules are explicitly mentioned in high school. The investigations in this unit can be used at all grade levels to address the standards. It is suggested to first do each simulation by hand and then move to technology in order to reinforce student understanding of the simulation process.

Several different applets and software can be used to carry out simulations. In addition to StatCrunch, mentioned in previous units, this unit makes use of TinkerPlots and Stapplet.com, as well as online simulation games. While the software introduced are useful, they are by no means the only software in existence that can be used to carry out these simulations. The examples shown in the investigations are meant to illustrate how a simulation can be set up to model the scenario; the simulation can then be carried out in any software that allows the user to do so.

References for This Unit

Martin, W.G. 2000. *Principles and standards for school mathematics.* Volume 1. National Council of Teachers of Mathematics.

National Governors Association. 2010. Common core state standards. Washington, DC.

UNIT 3B: Probability in Statistics

Now that we are familiar with the notion of probability and comfortable with the idea of simulations, we can begin to understand how these ideas are employed in statistics. Statistics utilizes randomness as a component of data collection in two ways: with **random sampling** and with **random assignment**. Imposing random selection in determining which units are included within a sample helps to reduce bias. **Bias** is the systematic favoring of certain units to be included within a sample. Random selection ideally produces a sample that is representative of the targeted population. That is, the distribution of the data in a sample is similar to the distribution of the data in the entire population. Random assignment of the units in a study helps to balance out the effects of potentially confounding variables and provides the foundation for establishing cause-and-effect relationships between variables. These are random processes through which we can collect data. Such random processes are foundational in statistics because they allow us to make decisions, inferences, and generalizations, and show cause and effect using data. *Inference* is being able to infer and generalize information using a sample to a population. *Cause and effect,* or causality, is making statements implying that one treatment or condition causes another—in other words, changes in values of one variable are a result of changes in values of another variable.

Random sampling is the pillar for statistical inference. For example, if a study is done with a group of volunteers, the results from that study may not be generalizable to the greater population. Suppose you are curious about whether people in the United States are in favor of allowing many more refugees to enter the United States on political asylum. If a poll surveys only those of refugee descent already living in the United States, the results cannot be generalized to the greater population because those of refugee descent are probably more likely to agree that many more refugees should be admitted than are U.S. residents in general. To better understand and generate accurate statistics regarding peoples' opinions about the crisis, a random sample of people should be polled, ideally resulting in a representative sample of the U.S. population.

Random assignment is the pillar for establishing causality. Consider Mr. Ditrick, a man in his early 60s who is trying to decide whether to retire. Mr. Ditrick decides to discuss the possibility of retiring with his friends, and one of them mentions that people who retire are more likely to die sooner than people who do not retire. To justify his friend's conclusion, Mr. Ditrick should seek a study that shows whether retirees are more likely to die sooner than those who do not retire. If a study wanted to make such a causal claim, in particular that retirement leads to an earlier death, the subjects of the study should be randomly assigned to one of two possible situations, either early retirement or retirement at a traditional age. Note that ethical considerations need to be made, which might limit the type of random-assignment experiments one could actually conduct. For example, people cannot be forced to retire early, and thus this study would be considered unethical. If such a study were possible, then the study would need to analyze if there were a significant difference in the proportion of deaths among people who retired early and the proportion of deaths among people who retired at a traditional age, so a causal claim could be made. However, when Mr. Ditrick asked his friend about his evidence, the friend stated that it had happened to many people he knew. He had observed that friends who'd retired early, died early, and those who retired later, died later. From this, the friend drew the conclusion that retirement causes early death. This friend was relying on anecdotal data. These anecdotal data do not take into consideration potential lurking variables, such as the fact that people may retire early because they are not well or that people who are retiring early might have the financial means for better health care. If one listens carefully, one may notice that people draw these types of conclusions all the time without considering the validity of the implications. Therefore, knowing about random assignment and understanding the role it plays in drawing conclusions is important in order to make valid decisions.

In this unit, part A and part B present case studies introducing the ideas of random sampling and random assignment found in current news articles. Additionally, the units provide investigations related to random sampling and random assignment and explore what probability and randomness afford us in statistics.

Unit 3B.a: Probability in Statistics: Random Sampling

We use probability in statistics in order to draw inferences—drawing conclusions about a population using only the information in a sample taken from that population. Random sampling provides us with the foundational conditions to make inferences. In this unit, we begin to describe random sampling and its importance to drawing inferences.

Suppose you are trying to gather information about a population but that population is so large that either (1) you cannot reach each individual in the population or (2) you do not have the resources to reach each individual. To help you get a sense about a characteristic(s) of the overall population, you might want to gather information from a subset of that population. In statistics, the subset of a population is known as a sample, and knowing the best way to select a sample is an important process. For example, suppose an administrator is interested in finding out students' opinions about whether the game Pokémon GO™ keeps students active and moving. If the administrator asks the opinion only of students who are currently on athletic teams, the administrator may conclude that, because of the beliefs of the students in the sample, everyone at the high school believes that Pokémon GO™ keeps them active and moving. However, because the students who were asked represent a distinct subset of all students at the school, the principal's conclusion about Pokémon GO™ keeping students active may not generalize to the entire high school.

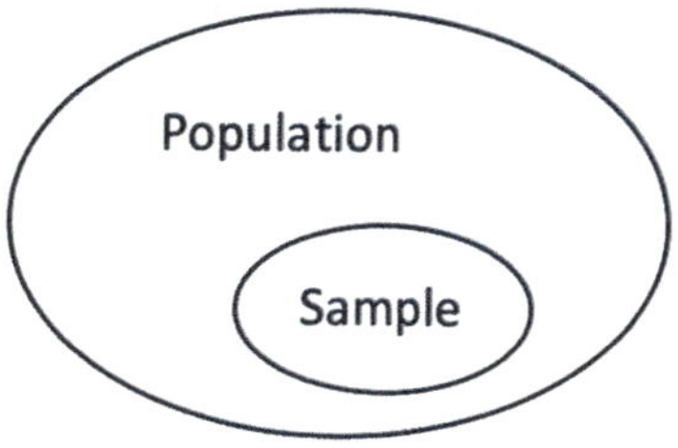

Case Study 6: Polling

We often want to use information from a sample to generalize and draw conclusions about the entire population. However, to do this accurately, the sample must have certain features aligned with those of the population. The selection of a sample can be done in many ways. For example, if the population of interest were adult voters in the United States, one could ask volunteers to take a survey about their political ideas, or one could obtain a list of registered voters in a certain party and survey people from that list. Consider the following case study article published in June 2016 on the Gallup website comparing Donald Trump's and Hillary Clinton's images (see www.gallup.com/poll/193043/trump-image-slips-clinton-holds-steady.aspx?g_source=ELECTION_2016&g_medium=topic&g_campaign=tiles).

The article discusses the fluctuations of net favorability of each candidate from August 2015 to June 2016. *Net favorability* is defined in the article as the percentage of people polled with favorable views minus the percentage of people with unfavorable views. The article discusses the net favorability of the candidates instead of the mere number of people with favorable views or the percentage of people with favorable views for each candidate in order to show how polarized the voter polls might be. The difference between favorable and unfavorable views provides an estimate for the difference in voter's opinions of each candidate.

The article states that surveys were collected "based on telephone interviews conducted June 13-19, 2016, on the Gallup U.S. Daily survey, with a random sample of 3,560 adults, aged 18 and older, living in all 50 U.S. states and the District of Columbia."

A **simple random** sample is a sample selected in such a way that each possible sample of a given size has the same chance (probability) of being selected. Thus, each of the possible samples is equally likely to be selected. Note that in simple random sampling, each individual in the population has the same chance (probability) of being selected as well. For the purpose of this book, we will say "random sample" to imply a simple random sample. Note that there are other types of strategies for selecting samples that utilize randomness that are beyond the scope of school-level curricula.

Suppose there were six people at a dinner party, but the dinner table was only set up for four people, so because of space, two people had to eat in the kitchen. The six people decided that they would randomly select two people who would then sit in the kitchen for dinner. To do this, they put their six names on small index cards, folded each card two times, and then put them in a hat. Then they mixed them and selected two names. The two names selected would go sit in the kitchen for dinner. This is an example of a simple random sample because each dinner guest had the same chance of sitting in the kitchen, and each possible sample of two people had the same chance of being selected. In this example, a nonrandom sample could be that the two people selected to go sit in the kitchen were the two people who were wearing the most colorful shoes. This selection would not be random because men's shoes tend to be less colorful than women's shoes; therefore, the women at the dinner party would be more likely to be selected to eat in the kitchen compared with the men at the party.

Having samples drawn in a random manner ensures that in the long run, **selection biases** (often also called sampling bias) are not present in the sample. Selection bias is one type of bias that favors a certain type of outcome due to the method of sampling. The Gallup poll concerning the 2016 presidential candidates states that 3,560 adults were selected in the random sample. If these adults were selected via a poll on a social-media site, for example, then this could result in a bias, because only certain people take the time to answer polls and surveys, and only certain people have access to the internet. This sample is not random from the population of all those adults that could vote because people are being asked to volunteer. This type of sampling would create systematic issues that would have been minimized if the sampling had been random from the entire population of voters. This may systematically result in outcomes favoring one particular candidate.

Another type of selection bias can be created when a sample is not representative of the population that one is interested in. Even though a sample may be randomly selected, it may still not be representative of the intended population. For example, if the 3,560 adults were all chosen randomly from a particular demographic, such as the very wealthy or a group of all women, sampling bias would be present because the wealthy might be inclined to favor Donald Trump and women might be inclined to favor Hillary Clinton. This may systematically result in outcomes to be in favor of one candidate. Although maybe the women were chosen randomly from the population of women, the article is interested in inferring results to the entire population of adults; thus, looking only within a particular demographic would bias the results and leave one unable to infer to the intended population.

It would not be in the best interest for a news article to publish results of a study if it had gathered information from only a particular demographic, because the results would not be inferable to the larger population, which in this case is voters in the United States. The purpose of the Gallup poll was to use the sample information to infer to the general voting population about the favorability of the two candidates. If the sample were biased, the study would not be able to do this.

Because random sampling minimizes sampling bias, this enables us to draw more reliable and representative inferences about the population. We do, however, have to be aware of the potential for bias being created by other things, such as nonresponse or response truthfulness. For example, if a survey was administered about the sexual activity happening at a school, students who have had bad experiences might be more likely to not answer the survey if mailed to them, or if interviewed, might be more likely to not give truthful answers. These situations would create a bias in the responses.

The Gallup article applies inference in the statement "For results based on the total sample of national adults, the margin of sampling error is ±2 percentage points at the 95% confidence level." The **margin of error** measures the variability of the statistic. If we were to sample repeatedly from the population, the margin of error measures the amount of expected variability of the resulting statistic when compared with the true value of the population parameter. This tells us that the results found from this sample can be inferred to the population with certain error caution. The margin of error allows using the descriptive summary values found from the sample to infer to the population by quantifying the potential variability due to sampling.

The last paragraph of the article discusses how the random sample was actually taken. It says, "Each sample of national adults includes a minimum quota of 60% cellphone respondents and 40% landline respondents, with additional minimum quotas by time zone within [the] region. Landline and cellular telephone numbers are selected using random-digit-dial methods." This is useful information because it allows us to understand where the data are coming from. To form the sample, a list of phone numbers was obtained. Forty percent of the time, a number was randomly selected from the landline, and 60% of the time, a number was randomly selected from the cell-phone list. This random-sampling process creates a random sample of 3,560 adults who were then surveyed by phone. Because the process was random, the sample minimizes sampling biases and mathematical theorems of inference can be applied to generalize the results to the population of all adult voters.

It's important to note that systematic sampling bias in samples is minimized when the sample is random. If sampling biases are present, then we are unable to generalize results to a population. In statistics, the importance of random sampling lies in the ability to draw inferences using samples to populations. With samples that have biases present, we are unable to infer results to the greater population.

CASE STUDY SUMMARY:

The main concepts developed in the polling case study are:

1. A simple random sample is a sample where each sample of the same size has the same probability of being selected.
2. Random sampling minimizes sampling bias in the outcomes due to the sampling method used and tends to provide a representative sample of the population.

Investigation 3B.a.1: Sampling of Words[5]

Goals of this investigation: Compute a sample statistic, receive an informal introduction to sampling distributions, and connect random sampling to probability and the ability to draw inferences.

The case study introduced the idea of random sampling as a way to avoid sampling bias when using sample information to infer to the larger population. In this investigation, we will explore the idea of random sampling and further explore what it affords us in statistics.

5 Activity adapted from Allan Rossmann and Beth Chance activity discussed in https://askgoodquestions.blog/2019/11/11/19-lincoln-and-mandela-part-1/.

A manuscript has been found with no author. Several historians wonder if Abraham Lincoln could have written the manuscript. They decide to have an analysis of a known manuscript he wrote, the Gettysburg Address, to judge if the writing style is similar. The length of the words in the Gettysburg Address is first investigated.

We begin by posing the following investigative question:

What is the typical length of a word spoken in the Gettysburg Address?

Consider the population of 268 words included in the Gettysburg Address given by President Abraham Lincoln in 1863:

> *Four score and seven years ago, our fathers brought forth upon this continent a new nation: conceived in liberty, and dedicated to the proposition that all men are created equal.*
>
> *Now we are engaged in a great civil war, testing whether that nation, or any nation so conceived and so dedicated, can long endure. We are met on a great battlefield of that war.*
>
> *We have come to dedicate a portion of that field as a final resting place for those who here gave their lives that that nation might live. It is altogether fitting and proper that we should do this.*
>
> *But, in a larger sense, we cannot dedicate, we cannot consecrate, we cannot hallow this ground. The brave men, living and dead, who struggled here have consecrated it, far above our poor power to add or detract. The world will little note, nor long remember, what we say here, but it can never forget what they did here.*
>
> *It is for us the living, rather, to be dedicated here to the unfinished work which they who fought here have thus far so nobly advanced. It is rather for us to be here dedicated to the great task remaining before us, that from these honored dead we take increased devotion to that cause for which they gave the last full measure of devotion, that we here highly resolve that these dead shall not have died in vain, that this nation, under God, shall have a new birth of freedom, and that government of the people, by the people, for the people, shall not perish from the earth.*

For the purpose of this investigation, we consider the words of the Gettysburg Address as the population and the variable of interest to be the length of the words in the address. To see what the average word length is, we could count the length of all the words in the speech and compute the mean (the population value of interest). Instead, as a statistical exercise, we can take a sample of words from the speech and if appropriate generalize the results from the sample to the population. It should be noted that in real-world circumstances, when we attempt to use sample information to generalize results to a larger population, we typically do not have access to the entire population. For our first sampling method, we will take our own sample of words using our eyes and show how to use the sample to estimate the length of the typical word in the population.

Begin by taking a sample of 10 words from the speech. To do this, take a pen or pencil and circle 10 words. After you have selected 10 words, continue reading the instructions.

Suppose, for example, we circled the following words:

liberty, civil, battlefield, dedicate, poor, nobly, honored, freedom, government, people

For each of our words, we count the number of letters in the word and record it.

Word	liberty	civil	battlefield	dedicate	poor	nobly	honored	freedom	government	people
# of Letters	7	5	11	8	4	5	7	7	10	6

The average length of words in this sample is seven letters. Is this a good estimate of the average length of words in the entire address?

To answer this question, we should ask ourselves whether the 10 words we circled are representative of the lengths of the 268 words in the population. Did we have some bias while we were sampling our words? Is our sample of words random? Let's investigate the answers to these questions. A class of teachers made two dotplots. The top dotplot shows the means from the teachers' 'by eye' samples. The bottom dotplot shows the means from random samples selected by software.

Dotplots of Random Selection and Self-Selection

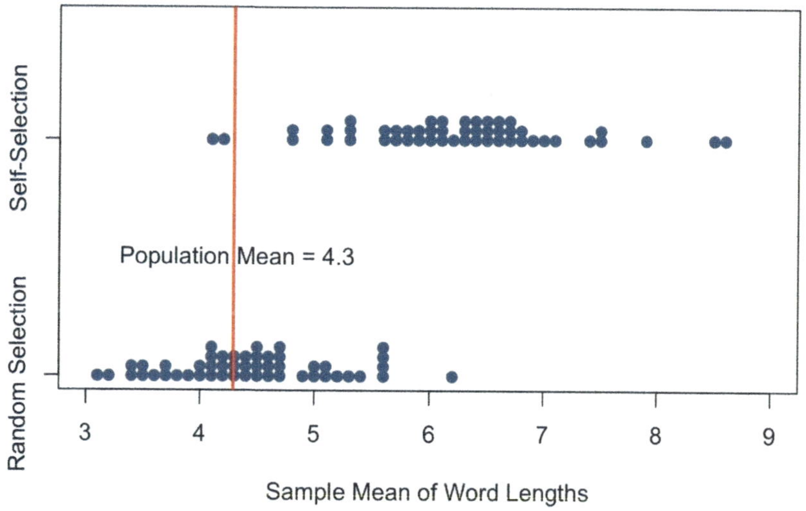

We see that the "by eye" self-selection method produced sample averages that are much higher than those that came from random samples selected by software. Our results show that our eyes gravitate toward the longer words, so instead of each word having equal probability of being selected for the sample, we have a size-bias and select the longer words with higher probability.

To further explore random sampling, we can place every word in the Gettysburg Address in a spreadsheet and then randomly sample 10 words. For example, the spreadsheet might look like the following image.

	A	B	C	D
1	Word	Length		
2	Four	4		
3	score	5		
4	and	3		
5	seven	5		
6	years	5		
7	ago	3		
8	our	3		
9	fathers	7		
10	brought	7		

To explore how the average number of letters in the Gettysburg Address varies depending on the sample chosen, we can have computer software generate random samples of size 10 and compute the average for each sample. In other words, if the random process of choosing 10 words and computing the average length of the words was repeated a large amount of times, how would the average behave? What is the probability of getting an average word length of seven as we did before?

Each dot on the following dotplot shows the average word length for one sample of size 10. There are 1000 dots because the software selected 1000 different samples of 10 words each.

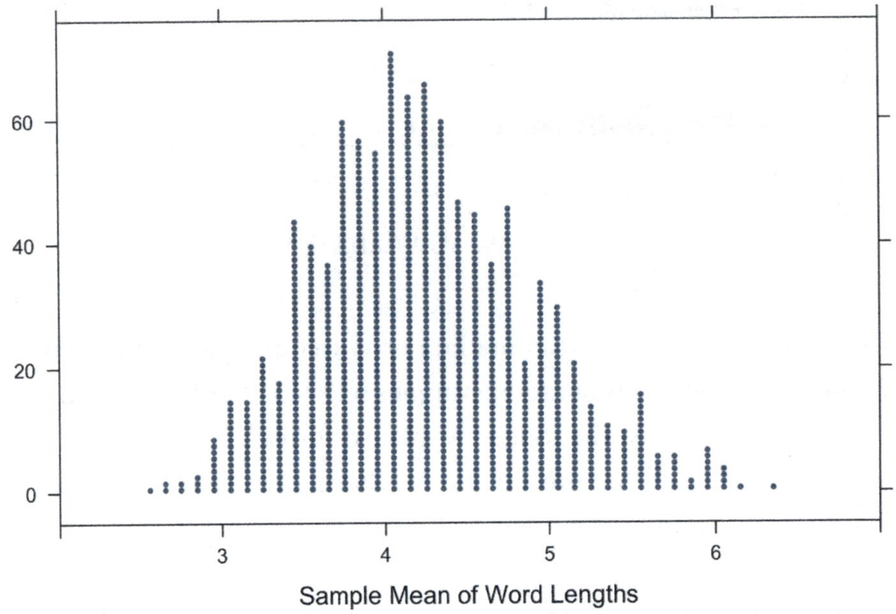

For example, there were four samples of size 10 that had an average word length of six letters. On the other hand, no sample had an average word length of seven. Does this surprise you? Why or why not? We can see from the dotplot that the average word lengths from random samples of size 10 are centered on 4.3. But there is a lot of variability in the averages from sample to sample. If additional samples of this same size are taken, we see that the dotplot is more clearly centered on 4.3. (See previous dotplot). This center is the mean of the means—the average of the sample averages. This indicates that, in the long run, the mean of the sample means settles around 4.3. This long-run connection to probability for the mean of the means will be further developed in the next unit.

It turns out that our initial sample of 10 words with an average word length of seven did not occur even one time in the 1000 samples that we simulated. In other words, the probability of getting a sample of 10 words with an average word length of seven was zero (or nearly zero). Of course, this probability was based on the fact that the samples were drawn at random.

Was the sample we took by circling 10 words drawn at random? The answer is no. Although we thought we selected random words, our eyes were not drawn to any of the short, seemingly insignificant words such as *we* or *a*. These short words were overlooked in our sampling. For our sample to have been truly random, we could have assigned each word in the Gettysburg Address a number—for example, one through 268—then picked 10 numbers between one and 268 out of a hat or using a random number generator. For the random sample in this table, the average word length is 4.3.

Random #	84	62	102	142	57	195	97	17	226
Word	gave	of	this	will	met	dedicated	proper	conceived	we
# of Letters	4	2	4	4	3	9	6	9	2

All of the words in the Gettysburg Address make up the entire population. So we can compare the pattern we see in the dotplot with the actual average word length in the entire population of words in the Gettysburg Address. The average population word length in the Gettysburg Address is 4.295. In the dotplot with 1000 samples, we see that many of the random samples selected had sample means around 4.3—the mean of the sample means is 4.3. It appears that the mean of the sample means is near the true population mean. This idea will be further explored in the next unit. This Gettysburg Address investigation shows us that we can see the pattern in the values of a sample mean value (in this case, the average word length) by repeatedly selecting random samples from a finite population.

Concepts of probability in the Gettysburg Address investigation are seen in two ways. First, they are seen in ensuring that each word in the address does in fact have the same chance of being selected (i.e., longer words do not have a greater chance of being selected). Second, the notion of probability as the long-term outcome of a random process is touched upon in this example. In the long run, the probability of the mean of the means, 4.3, settles close to the true population average word length.

INVESTIGATION SUMMARY:

The main concepts developed in the sampling of words investigation are:

1. A sample statistic can vary from sample to sample. A dotplot can be used to visualize the distribution of the statistic from sample to sample.
2. In simple random sampling, each outcome is ensured the same probability of being selected. Without random sampling, then our results may not be valid for drawing conclusions about the population.

Follow-Up Questions

1. AngieClothing, a women's clothing store, wants to know how many times in a year women in the United States shop at its stores. Of the women who subscribe to AngieClothing emails, 1,500 are sampled, and 80% of them reported shopping at AngieClothing four to six times a year. Is 80% likely to be a realistic estimate for the true percent of women in the United States who shop at AngieClothing four to six times a year? Why or why not?
2. A state initiative came out requiring all public high schools to give an assembly on the dangers of driving while texting. After the assembly, pledge cards were passed around so students could vow not to text and drive. To see if the program worked, a random sample of 1000 students who sent in cards saying they intended to quit texting while driving were contacted three months later. It turned out that 130 of the 1000 sampled individuals had not texted over the past three months. What is the population? What is the sample? Based on these data, can one conclude that the initiative was a success? Why or why not?
3. Suppose an administrator of a large district is curious how many teachers work over summer break. He does not have time to ask all of the teachers in the district about their summer plans, so he would like to contact a sample.
   a. How would you suggest the administrator do this?
   b. Is it important for the sample to be random? Why or why not?

Unit 3B.b: Probability: Random Assignment

The first section of this unit focused on the foundational ideas of making inferential statements. Another important notion in statistics is that of making causal statements. Making causal statements is a goal of many statistical studies. We design many studies and experiments to understand how one variable can influence another. For example, someone may design a study to investigate if receiving tutoring in mathematics can cause students to get better grades, or whether taking a specific drug might cure a patient of a certain disease.

To investigate such questions, subjects participating in the study must be randomly assigned to different conditions. In the case of mathematics tutoring, the subjects may be high-school students, and they may be randomly assigned to two conditions; one condition would be receiving tutoring and the other condition would be not receiving it. Then, grades of students receiving tutoring could be compared with grades of students not receiving tutoring. For a study on the effectiveness of a drug, a group of patients with a certain disease could be the subjects and they could be randomly assigned to either receiving a drug or not receiving a drug. The effects of the drug could then be examined by looking at how the patients' disease progressed for those patients receiving the drug versus those not receiving the drug.

Suppose you are trying to understand how a type of behavior or an intervention affects another. You are interested in exploring the cause-and-effect relationship between two actions. To make cause-and-effect conclusions, one might conduct an experimental study. Remember that you explored the differences between observational and experimental studies in Unit 1. An **experimental study,** or an experiment, is one where the researchers manipulate some condition in the study. For this type of study, researchers do not merely observe to collect data. Instead they implement a treatment, an intervention, a procedure, or a program and then observe the result. Contrary to observational studies, in an **experimental study,** researchers impose a treatment and an outcome is measured. In experimental studies, treatments are applied to the experimental units, and then the units are measured on some desired outcome. Experimental studies allow researchers to study whether changes in the treatment resulted in corresponding changes in the response variable. To make cause-and-effect statements, the observational units must be **randomly assigned** to different **treatments** or **conditions**. A treatment or condition in an experimental study can be defined as the factor that is being studied in the experiment. For example, suppose researchers are interested in understanding whether an online mathematics-tutoring program is as good at increasing student achievement as in-person tutoring. To conduct an experiment, researchers may recruit a group of students in a school and then **randomly assign** them to receive in-person mathematics tutoring or online mathematics tutoring. The treatment is the type of tutoring received, either the in-person tutoring or the online tutoring.

Random assignment implies that a random process, such as flipping a coin, is used to decide whether a subject would be put in the in-person or online tutoring. Because this study analyzes student achievement, it is important for the students to be randomly assigned to the two types of tutoring in order to ensure that each group of students does not have some specific set of characteristics. For example, if students were allowed to choose their own tutoring program, then it is possible that the students who would select online tutoring would be those that generally spend a lot of their time on computers. This could lead to bias in the study, because online tutoring might be good at raising achievement only for students who are comfortable using a computer. Randomly assigning students to the two groups mitigates the influence of outside factors, such as background computer knowledge, because presumably both treatment groups will have students with a range of computer experience. For example, consider a weight loss study with 100 volunteers. Researchers randomly assign the group to two conditions—one to the new diet and one to their current diet. Several factors could influence the amount of weight loss a person could experience, namely, how much they exercise or how healthy they normally eat. With random assignment, these factors should be equally distributed across the treatment and control diets. In other words, the distribution of exercise amounts across the two conditions should be similar. That goes for all potential influencing factors. In an experiment, the researcher randomly assigns which groups receive which treatments. The goal of random assignment is to balance the effects of unmeasured variables across groups.

Case Study 7: Flossing

To be careful consumers of data, one must consider the value of random assignment in an experiment as it allows one to discuss cause and effect. Consider this *New York Times* article about the effects of flossing on preventing cavities and gum disease: www.nytimes.com/2016/08/03/health/flossing-teeth-cavities.html?mwrsm=Email&_r=1.

The article discusses how, although it has been suggested by both dentists and the federal government in the dietary guidelines provided by the Department of Agriculture and Health and Human Services that flossing will prevent cavities and gum disease, this cause-and-effect relationship has not been appropriately researched. The article mentions that among experts, it has been an "open secret" that flossing has not been shown to cause a reduction in cavities or gum disease.

The article mentions 12 randomly controlled trials that examined the effect of flossing on plaque reduction after one to three months that were published in the Cochrane Database of Systematic Reviews. **"Randomly controlled studies"** tell us that in these 12 studies, people were randomly assigned to either the condition of flossing or not flossing for a period of three months. In a study of this type, it is important to randomly assign participants to a flossing group and others to a non-flossing group due to potential confounding variables that might arise. For example, levers of gum disease among participants at the beginning of the study. A researcher would not want all participants with severe gum disease to be assigned to the same treatment group. Similarly, for other levels of gum disease. A researcher relies on random assignment to balance out the various levels of gum disease between the two groups (flossers and non-flossers). At the one-month mark and the three-month mark, the participants' teeth were examined to see the amount of plaque present. The article continues by noting that flossing has been shown to have some benefit for gum disease, or gingivitis, but notes that the evidence was low. Stating that the evidence was low might imply that the difference between the two groups, the flossers and non-flossers, was not that large, or possibly that some of the studies found no difference between the two groups and some found only slight differences between them. Although there might have been some difference in participants' gums, there was not a drastic difference.

Later the article says that brushing with fluoride has been proven to "prevent dental decay." This implies a cause-and-effect conclusion, which implies that random assignment was used. People were assigned to the fluoride and nonfluoride treatments for an extended period of time, and their dental decay was measured at different points. The difference in the amount of dental decay between the nonfluoride group and the fluoride group was large, thus providing strong conclusive evidence of the cause-and-effect relationship between fluoride treatment and the reduction of decay.

Although a connection between fluoride and decay has been found, the relationship between fluoride and gum disease or cavities has not been found to be statistically significant. This is addressed in the article with an interview with one of the doctors. Dr. Sebastian G. Ciancio's comment suggests that longer randomized studies should be used to further test the effect of fluoride on gum disease and cavities. From the studies that have been conducted, one may not see effects in the short term, but it is possible that a long-term study may lead researchers to see the effects 20 years down the line. One challenge of a randomized experiment over a long period of time is that it might not be realistic to carry out. Thus, despite the lack of rigorous studies, dentists still recommended that people floss.

This case study illustrates some of the difficulties with conducting randomized experiments. Although they are the gold standard and necessary to claim cause and effect, they are at times very difficult to carry out in real life. In statistics, the importance of random assignment lies in the ability to make cause-and-effect claims.

CASE STUDY SUMMARY:

The main concepts developed in the flossing case study are:

1. An experimental study, or an experiment, is one where the researchers manipulate or impose some condition in the study.
2. Experimental studies allow researchers to study cause and effect. To make causal statements, the observational units should be randomly assigned to different treatments.
3. Random assignment implies that a chance procedure is used to assign subjects to a treatment in the study.

Investigation 3B.b.1: Swimming With Dolphins[6]

Goals of this investigation: Illustrate how knowing the long-run probabilistic behavior of a statistic can help determine whether the statistic happened by chance or was caused by the treatments in the experiment.

Researchers published a paper exploring the effects of swimming with dolphins as a treatment for depression. Specifically, they aimed to answer the following investigative question:

Does swimming with dolphins help alleviate signs of depression?

Their paper, published in the *British Medical Journal* in 2005 (Antonioli and Reveley, 2005) and found at http://archive.is/I3Bow, discusses an experiment that randomly assigned 30 patients who were diagnosed as depressed to two groups—one group of 15 patients was designated to swim with dolphins and the other group of 15 patients was designated to not swim with dolphins. As discussed, the reasoning behind randomly assigning units to the two groups is to help mitigate and balance the effects of **confounding variables**. A confounding variable is a variable that affects the variables being studied, thus potentially masking the actual relationship between the variables under investigation. For example, hot chocolate sales and tissue box sales might appear related because they both increase and decrease at the same time; however, the weather temperature is a confounding variable. Both the hot chocolate sales and the tissue box sales are related to the temperature, but not to each other. Therefore, the temperature is masking the actual relationship between the variables. In this investigation, the researchers used an experiment because they wanted to make a causal statement regarding depression and swimming with dolphins.

To answer the investigative question, we need to look at the difference in the number of people who showed a decrease in symptoms of depression in the dolphin group versus the number of people who showed a decrease in symptoms in the nondolphin group. We want to consider the difference between the two groups' symptom improvements and try to understand whether that difference could have happened by the chance variation resulting from the random assignment of participant to treatment groups, or whether the difference was in fact a result of swimming with the dolphins. To do this, we look for patterns in the long-run behavior of the difference of the two

6 Activity adapted from Strayer and Matuszewski (2016).

groups. This is where probability comes into play. To aid our investigation, consider these guiding questions:

How often can our observed difference of symptoms occur if we assume random chance?

What is a typical difference that could happen just by random chance?

The researchers' results indicated that 13 of the 30 participants showed improvement in their depression level at the completion of the study. Assuming that swimming with dolphins had no effect on depression levels, of the 13 participants that improved, how many would you guess would be in the dolphin group? One way to approach this question is to understand what would happen if the dolphins had no effect on the depression levels.

Let's suppose that after either swimming with dolphins or not swimming with dolphins, there was no difference in the number of people that showed improvement between those swimming with dolphins and those not. If this were true, then in the long run, this would indicate that if the experiment were repeated over and over again, we would expect the average difference in the number of people that showed improvement between those swimming with dolphins and those not to be close to zero, evidence that there was no difference between the two groups. Another way to think about the difference being zero is that each group, those swimming with dolphins and those not swimming with dolphins, would have about the same amount of people who showed improvement. No difference would indicate that out of the 13 total improvers, we would expect there to be about six to seven improvers, or about half the 13 total improvers, in each of the two groups, as pictured in the following image. We would expect this to be about half of 13 because the groups are the same size. If the groups were different sizes, it would not be half. This idea will be discussed further when designing a simulation for this problem.

However, in the actual study, 10 of the 13 improved patients were from the dolphin group. So, the difference in the number showing improvement between the two groups is seven (10-3). Can this difference be explained by chance variation alone? *If there had*

been no difference between the two groups, would this be a reasonable result? In other words, could these 10 patients have shown improvement in the short run when, in the long run, if there was no difference in depression levels, we would expect to see about six or seven people improve in each group? How likely is it that 10 of the 13 improvers randomly fall into the dolphin group if we assume that swimming with dolphins is the same as not swimming with dolphins for depression? Essentially, we are asking whether it is possible for the observed difference of seven is a reasonable outcome from the random assignment of participants to the two groups if swimming with dolphins has no effect on treating depression.

To gain some intuition about these likelihoods and the long-term patterns of the difference, we can design a simulation for the experiment, perform the simulation a large number of times, record the differences resulting from each trial, and determine which difference are likely and which are not likely. We then can use the information from the simulation to determine if the observed difference in the study (10 improvers in the dolphin group and 3 improvers in the nondolphin group) is very unlikely. If the observed difference is unlikely, then we can conclude that swimming with dolphins can in fact be attributed to improving the depression levels of patients.

One way to simulate this experiment is described below:

Each of the 13 improvers will be randomly assigned to one of the two groups by flipping a coin and seeing whether they are in the dolphin group or not. If the coin lands heads, then the improver is in the dolphin group. If it lands tails, then the improver is in the nondolphin group. Note that we expect that in the long run six or seven (6.5) should be assigned to the dolphin group. After assigning the improvers to either group, we can then tally up the total number of improvers in each group. We would repeat this 1000 times to find what is expected to happen in the long run. We could do this by hand and flip a coin, but doing this amount of coin flips would be too time-consuming and tedious, so we employ the use of software instead to carry out our simulation. It is important to note that this simulation using a coin works because the two groups, dolphin and not dolphin, are the same size (15 people in each group). If instead there were unequal amounts of people in each group, then we could no longer use a coin to simulate; instead, we would need to simulate using manipulatives such as cards. Simulating with cards would entail designating red cards to represent those that remained depressed and black cards to those that improved. Then instead of flipping a coin, we would be shuffling and redealing.

The following histogram illustrates how many of the 13 improvers land in the dolphin group when they are randomly assigned. We can see that if we flip a fair coin (which gives an equal probability of an improver being in the dolphin group or nondolphin group), then we can see by the histogram of the counts that the number of improvers we would expect to see in the dolphin group would be around six or seven. This is demonstrated through the histogram because the six and seven bars have the highest frequency and the balance point of the histogram is around the mean of 6.5, as indicated by the fulcrum in the histogram. Looking at the histogram, what was the frequency of getting 10 improvers or more in the dolphin group?

We can see that in 1000 repetitions of this simulation, approximately 35 times (or 0.035 percent of the time), there were 10 improvers or more in the dolphin group, as the researchers found.

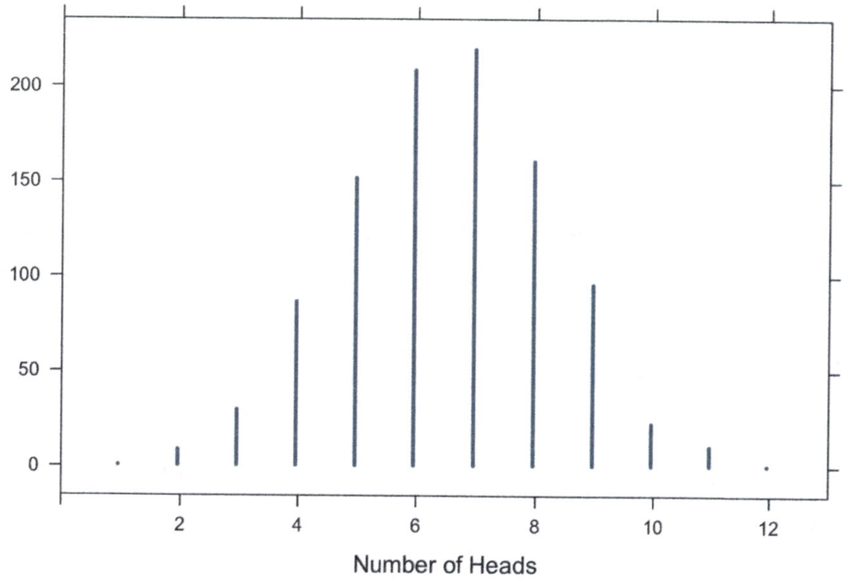

This indicates that the result of getting 10 improvers randomly in the dolphin group is quite unusual. In other words, this is strong evidence that getting 10 improvers in the dolphin group would not happen by chance, indicating that participants in the dolphin group actually improved. In fact, there is strong conclusive evidence that the participants' depression decreased because the dolphins did have an effect. These ideas are formally discussed in randomization tests, *p*-values, and hypothesis tests found in the AP statistics curriculum.

A description of how to carry out this example by hand using a deck of cards is shown by Strayer and Matuszewski (2016). The full article can be found at www.nctm.org/Publications/Mathematics-Teacher/2016/Vol109/Issue8Statistical-Literacy_-Simulations-with-Dolphins/.

This example illustrates another connection of how probability is used in statistics. Knowing the long-run probabilistic behavior of the number of improvers being in the dolphin group allows us to ascertain that the study result is in fact not due to chance. However, we must be reminded that, in practice, we carry out an experiment only one time. We do not repeat an experiment over and over again and assign participants to treatments multiple times. The purpose of both this simulation and considering long-run probabilistic behavior is to apply the same statistical understanding to a single experiment and to use the results from that experiment to determine whether an observed difference between two treatment groups can or cannot happen just by chance. If it is unusual for the results to happen by chance, then we have a rationale for establishing a causal relationship,e.g. that swimming with dolphins caused improvement in depression levels.

INVESTIGATION SUMMARY:

The main concepts developed in the swimming with dolphins investigation are:

1. If we know the long-run behavior of how the difference between two groups behave, then we can judge if the difference happened by chance or if the treatment caused the difference.
2. The notion of probability as a long-run relative frequency can be applied to the context of cause and effect.
3. Random assignment is used to balance out the effects of potential confounding variables.

References

Rossmann, A. 2019. #19 Lincoln and Mandela, part 1. *Ask Good Questions*, November 11. https://askgoodquestions.blog/2019/11/11/19-lincoln-and-mandela-part-1/.

Strayer, J., and A. Matuszewski. 2016. Statistical literacy: Simulations with dolphins. *Mathematics Teacher*, 109(8): 606–11.

UNIT 3C:
Sampling Distributions and Bootstrapping

The prior unit focused on how probability is connected to statistics. In this unit, we will make this connection even more explicit. Oftentimes in statistics we are presented with situations where we do not have access to the entire population; however, we want to know something about certain characteristics of the population as defined by either quantitative or categorical variables. We summarize these variables with parameters such as the population mean or population proportion. We likely have access only to a sample of the population, and we need to use only that sample to draw conclusions about the entire population. While an important goal in statistics is to make claims and provide information about the population, to do so, we must somehow be able to connect the information we gather from the sample to the population.

There are two main approaches to this inference problem that are important to consider. The first relies on classical methods of building confidence intervals using the normal distribution, and the second relies on modern and technologically heavy bootstrap methods. In this unit, we will discuss both methods, in order to give a full picture of the ways modern statistics approaches inference. We will begin with building the foundation for inferential methods using the normal distribution and deriving the central limit theorem, and conclude with building the foundation for inferential methods using the bootstrap method.

For both the traditional approach and the bootstrap approach, there are two important steps to connect the sample to the population: (1) selecting or having access to a *random* sample of the population, and (2) using the information from the random sample to draw inferences about the population. In past investigations, we have repeatedly selected multiple random samples from a population in order to develop theory and understanding, but in practice we select only **one** sample to draw inferences. This process is depicted as follows:

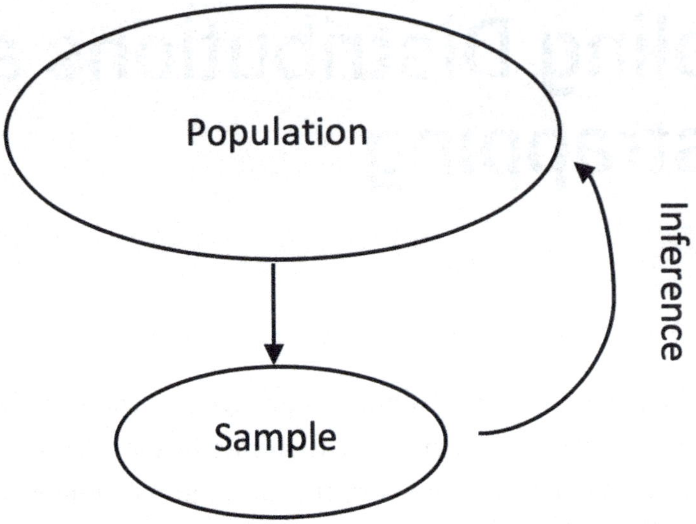

More specifically, we are interested in knowing information about a **population parameter**. A parameter is a characteristic about the population that can be quantified in some way, such as the population mean for a quantitative variable or the population proportion in a particular category for a categorical variable. We use the matching **sample statistic** (e.g., sample mean or sample proportion) to help us draw inferences about the population parameter of interest. Therefore, a parameter is at the population level, and a statistic is at the sample level.

Sampling variability describes the sample-to-sample variation of a sample statistic. Suppose, for example, one took a sample from a population and computed the mean of that sample. It is important to understand that if a different sample had been selected, then the calculated mean for that sample would more than likely not be the same. For example, in the prior units, using simulations we constructed a dotplot to represent the simulated distribution of the sample. This dotplot, pictured again below, shows that the sample mean word length varies from sample to sample (each dot represents a sample mean). This variation in the values of the means is called the sampling variability of the sample mean.

Note that when random samples of the same size are repeatedly selected from a population, sample statistics such as the sample mean or the sample proportion vary from one sample to another. This sample-to-sample variability is called the sampling variability of the statistic. The sampling distribution of a statistic describes this sample-to-sample variability.

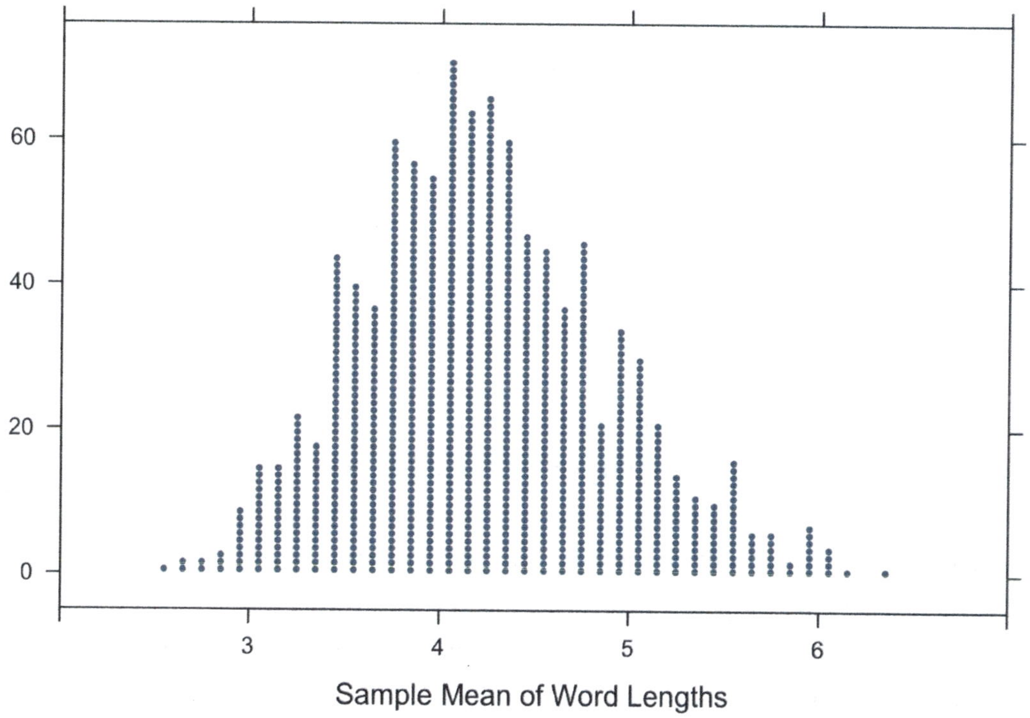

While the values of a sample mean are not necessarily the same from sample to sample, certain values may be more likely to occur than others. In fact, the distribution for some sample statistics follows a specific pattern, which we will discover. A distribution of a sample statistic is called a **sampling distribution**.

Up to this point, we have covered two different types of distributions: population distributions and sample (or data) distributions. In this unit, we are focusing on *sampling distributions.* A sampling distribution is constructed by taking all possible samples of the same size from a population and then computing and recording the value of the sample statistic for every possible sample. To construct the entire sampling distribution, values of the sample statistics need to be computed for *all* possible samples. If all samples are not represented, then we call the created distribution an **approximate sampling distribution**. We can obtain an approximate sampling distribution through simulations. The dotplot from the Gettysburg Address sample means provides an example of an approximate sampling distribution. Even though the approximate sampling distribution does not include the sample means from all possible samples, it still provides information about which values of the statistics are surprising and which values are common. An approximate sampling distribution thus provides a structure for us to observe common sample statistics and surprising sample statistics.

In general, a sampling distribution provides information on the possible values of a statistic and how often those values can occur. Knowing the sampling distribution thus enables us to draw inferences about a population based on data in a single sample. If we had access to only one sample statistic but knew information about how surprising or common our statistic might be, then we could estimate what the value of the population parameter might be.

Sampling distributions enable us to make connections between a sample and a population. The connection relies on knowing important features of the sampling distribution. Can we predict the shape of the sampling distribution for a statistic? Can we predict the center and the variability of the sampling distribution?

This connection is difficult to grasp, but the investigations in this unit will walk through this process step-by-step. The goal of this unit is to learn about sampling variability, examine the sampling distributions of commonly computed sample statistics, and understand how to use sampling distributions to help us make inferences about a population. For clarity, throughout the unit we will focus on the sample mean and its sampling distribution.

We begin by providing some background information on the theoretical normal distribution, which will be important in this unit. The normal distribution is a symmetric, single-mounded, bell-shaped distribution that has specific properties and is often found in nature and used in statistics. The Normal distribution is completely specified by its mean and its standard deviation. A graphical display of the distribution follows:

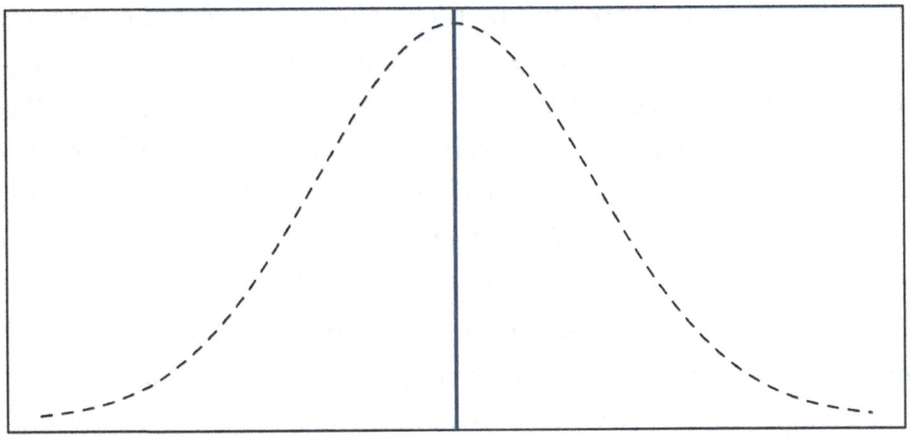

In a normal distribution, the area under the curve represents probability. For example, the probability of a value falling between a and b is represented by the shaded area as shown below:[7].

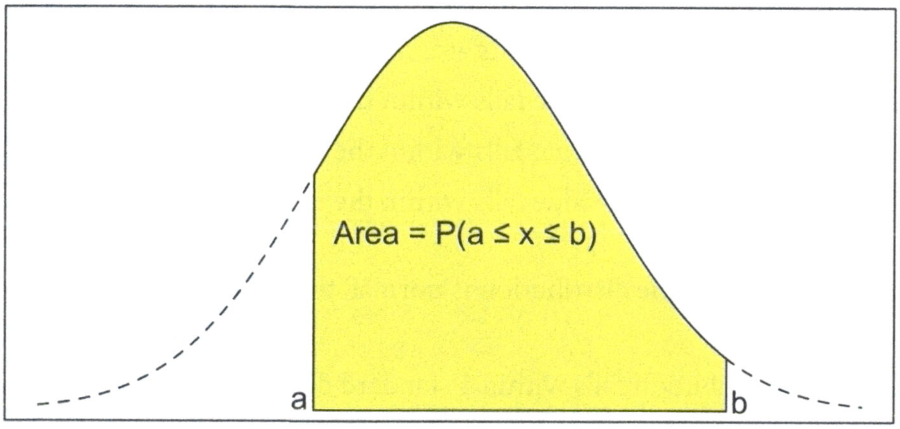

The mean of the normal distribution is denoted as μ and represents the center of the bell. The standard deviation is denoted as σ, sigma, and represents the average distance from the mean. A bell-shaped, symmetric distribution has some specific properties that relate the mean, the standard deviation, and probability. These properties are often referred to as the empirical rule. Namely:

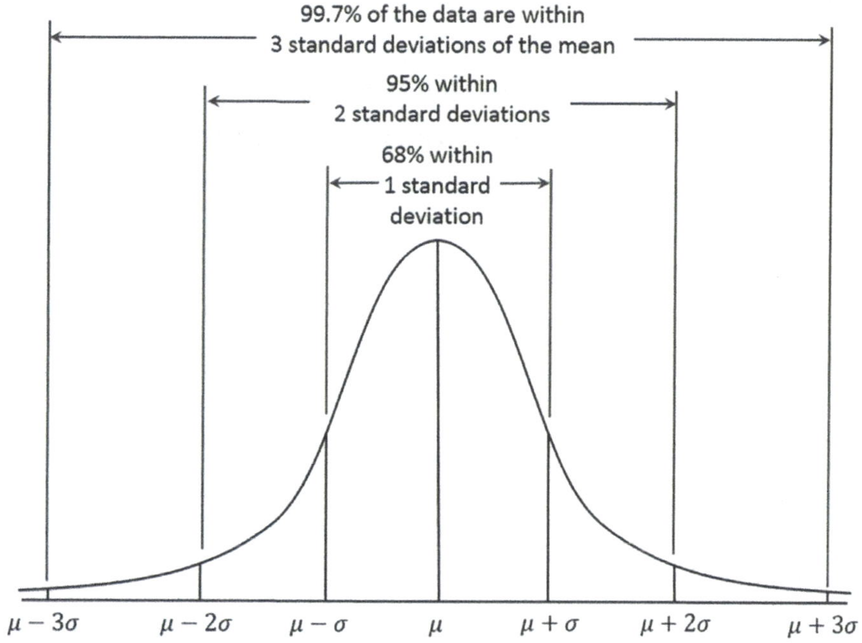

Graphic taken from https://commons.wikimedia.org/wiki/File:Empirical_Rule.PNG. Original created by Dan Kernler, CC BY-SA-4.0, via Wikimedia Commons under the Creative Commons license.

7 Note that the area under the curve between a and b is the same as the area under the curve between a and b including a and b. This is because the area under a curve at one point is zero. In other words, $P(a<x<b) = P(a \leq x \leq b)$.

- ~68% of the distribution falls within 1 standard deviation of the mean,
- ~95% of the distribution falls within 2 standard deviations of the mean, and
- ~99% of the distribution falls within 3 standard deviations of the mean.

Written mathematically, these statements are:

- there is a 68% chance the value falls within the interval $(\mu - \sigma, \mu + \sigma)$,
- there is a 95% chance the value falls within the interval $(\mu - 2\sigma, \mu + 2\sigma)$, and
- there is a 99% chance the value falls within the interval $(\mu - 3\sigma, \mu + 3\sigma)$.

If the bell-shaped, symmetric distribution is normal, these properties are even more specific. They are:

- 68% of the distribution falls within 1 standard deviation of the mean,
- 95% of the distribution falls within 1.96 standard deviations of the mean, and
- 99% of the distribution falls within 2.576 standard deviations of the mean.

Using the information of the normal distribution provided, we now embark on our first investigation.

Investigation 3C.1: Sampling of Words, Part 2

Goal of this investigation: Introduce sample-to-sample variation of a statistic and construct an approximate sampling distribution for a statistic through simulation.

In this investigation, we will construct an approximate sampling distribution by taking multiple random samples from a population and use it to determine which values of a statistic are common and which ones are uncommon. Consider the following investigative question:

What are common and uncommon values of the mean length of words in samples of size 10 from the Gettysburg Address?

In the Gettysburg Address investigation, we simulated the process of taking samples of size 10 and computing the average word length of the 10 words selected from 1000 samples. Then we plotted the following dotplot, where each dot represents the average word length of the words in one sample of size 10 from the 1000 random samples.

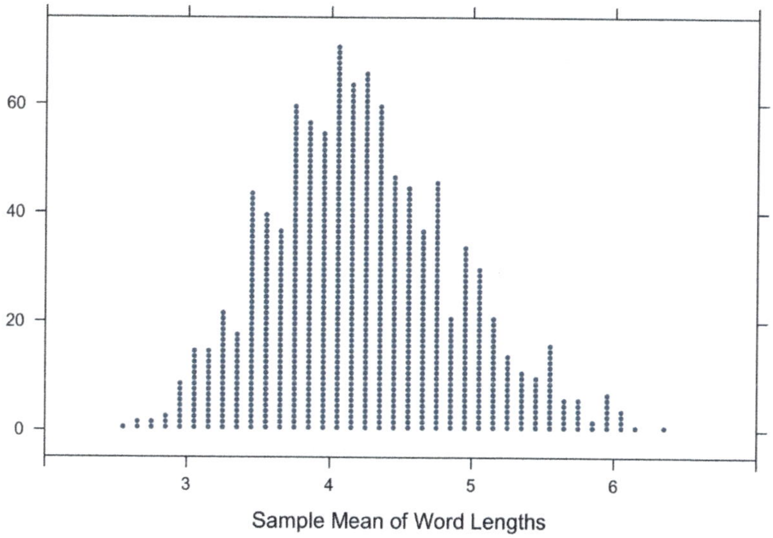

Approximate sampling distribution for mean word length of samples of size 10

From the dotplot, we can see that many samples had an average word length between 3.5 to 5.5 letters. Less common values were around the average of three letters or less and six letters or more. The dotplot also shows us that out of the 1000 random samples of size 10, none of them had an average word length of seven or above, as well as two or below. These observations from the approximate sampling distribution provide us with a general picture of the behavior of the sample mean, which, in this case, is the statistic that connects to our investigative question. Although the approximate sampling distribution as visualized in the dotplot does not represent the entire sampling distribution, it does show us information about which values of the sample mean (our sample statistic) are surprising and which values are common.

It is important to recognize that sampling distributions show both the possible values of a statistic and how often these values could occur if random samples are repeatedly selected for the population. In addition, one should recognize that values of statistics that occur often are common or plausible and those that are far from the values that occur often are considered surprising or uncommon.

As stated, the Gettysburg Address dotplot depicts an approximate sampling distribution. Up to this point in the book, we have discussed three different types of distributions:

1. population distributions,
2. sample (or data) distributions, and
3. sampling distributions.

To ensure we are clear on the differences and connections among these three distributions, we answer the following investigative question:

What are the connections among the population distribution, sample (or data) distribution, and sampling distribution?

In the introduction to this unit, we distinguished between a population distribution, a sample distribution, and a sampling distribution. Let's make sure we understand the distinction between these three distributions in the context of this problem.

In the GettysburgAddressWordwLengths.csv file, we have the words of the Gettysburg Address. The population is the entire list of words in the address. The following dotplot illustrates the *population distribution* for this data set:

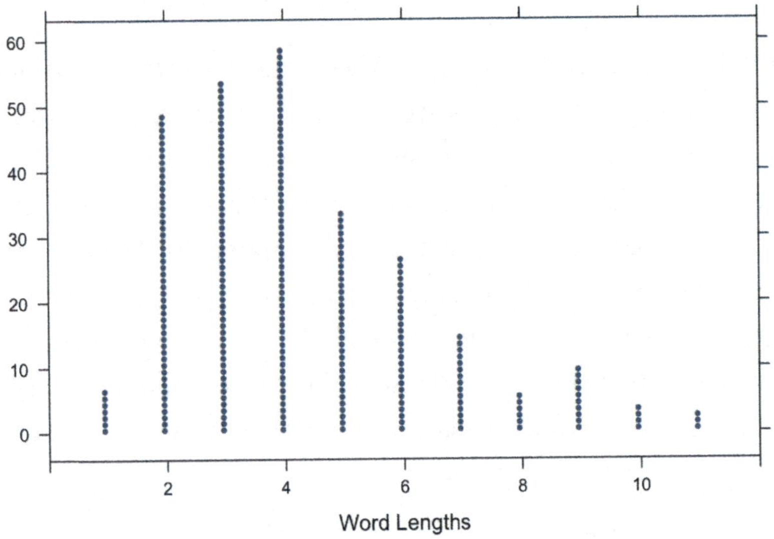

Population distribution of word length

The mean word length and the standard deviation of the word length of this population are 4.29 and 2.12, respectively. Note that a value for word-length in the population must be a whole number, while the population mean and the standard deviations do not (in this case 4.29 and 2.12 are both decimal values).

Next, let's create a possible sample distribution by taking one random sample of size 10 and plotting the length of each word from the sample. The words that are randomly selected by software are *and, men, birth, dead, they, shall, nation, we, a,* and *brave.*

This *sample distribution* can be visualized in the following dotplot:

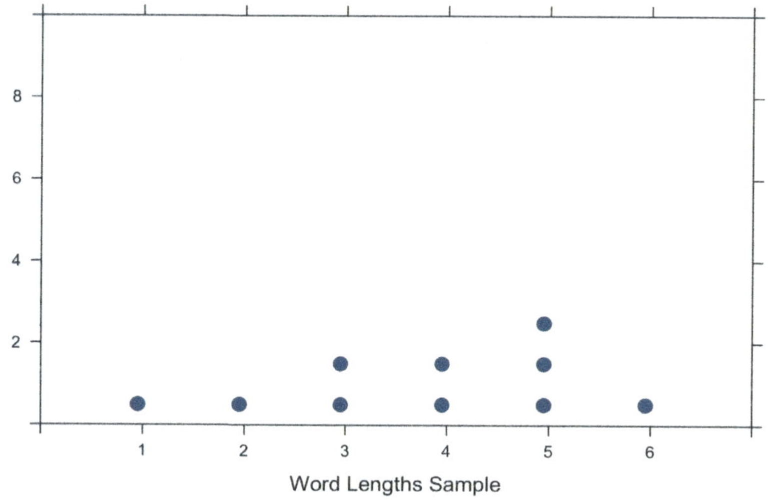

Sample distribution for word length of a sample of size 10

For this sample, the sample mean is 3.8 and the standard deviation is 1.55. We would expect that if another sample were selected, then the sample distribution, the sample mean and sample standard deviation may be different. However, while the sample distribution may change depending on the sample selected, the population distribution always remains the same. This is because it is the distribution for the entire population and thus does not vary.

As described in the initial part of this investigation, the following dotplot depicts an approximate *sampling distribution*:

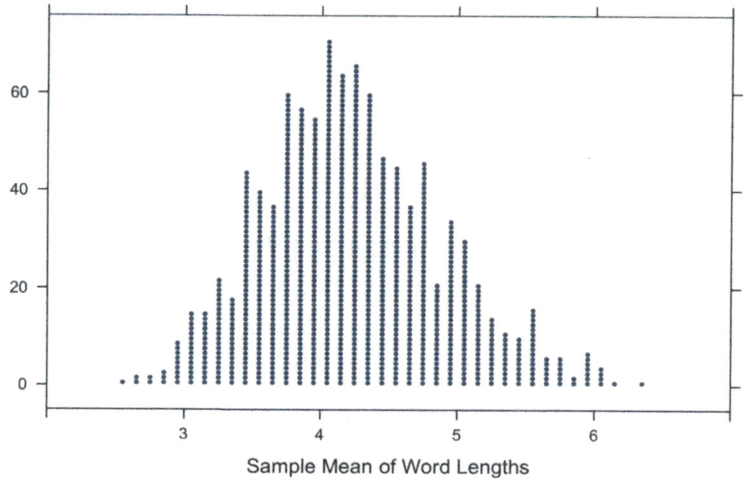

Approximate sampling distribution for mean word length of samples of size 10

The values in this dotplot are the sample means found from taking 1000 repeated samples from the population of words. The mean of this sampling distribution is 4.28 and the standard deviation is 0.65.

The three dotplots represent the three types of distributions (population distribution, sample distribution, and sampling distribution). The next investigation will further illustrate the relationship between the population distribution, the sample distribution, and the sampling distribution.

Frequent difficulties in understanding sampling distributions lie in differentiating among the population distribution, a sample distribution, and the sampling distribution of a statistic. The population and sample distributions describe the possible values of a variable and how often each value occurs within a population or sample, respectively. The sampling distribution, introduced in this unit, describes the sample-to-sample variation in possible values of a *statistic* from multiple samples of the same size sampled from the same population. It is essential that one is comfortable with these three different types of distributions in order to proceed to inference.

INVESTIGATION SUMMARY

The main concepts developed in the sampling of words, part 2 investigation are:

1. There are three different important distributions: population distribution, sample distribution, and the sampling distribution.
2. The sampling distribution is a distribution that describes how a statistic varies for repeated samples from the same population.
3. The sampling distribution may show which values of the statistics are common or plausible and which are surprising or unusual.

In the next investigation, we examine the relationship between the shape of the population distribution and the shape of the sampling distribution, as well as the effect of the sample size on the sampling distribution.

Investigation 3C.2: Different Pedagogies

Goals of this investigation: Informally discover the shape of the sampling distribution and examine the effect of the sample size on the shape of the sampling distribution.

In this investigation, we are going to examine the predictability of the shape of the sampling distribution for a sample mean. To do this, we must first investigate its shape. Our investigative question is:

Can the sampling distribution for the sample mean be modeled by the normal curve regardless of the shape of the population distribution?

A school has been experimenting with different pedagogies for teaching mathematics. The different strategies are teaching traditionally, teaching technology-based, teaching self-paced, and teaching project-based. There were a total of 4316 students, with 1079 students exposed to each pedagogy throughout the year. At the end of the year, all of the students took a mathematics test. Pedagogy_Methods.csv contains the test results. Using these data, let's investigate the relationship between the shape of the population distribution and the shape of the sampling distribution.

We begin by visualizing the population distributions of student test scores for the students in each of the four pedagogies. These population distributions describe the test scores for all 1079 students after being taught by the four teaching strategies.

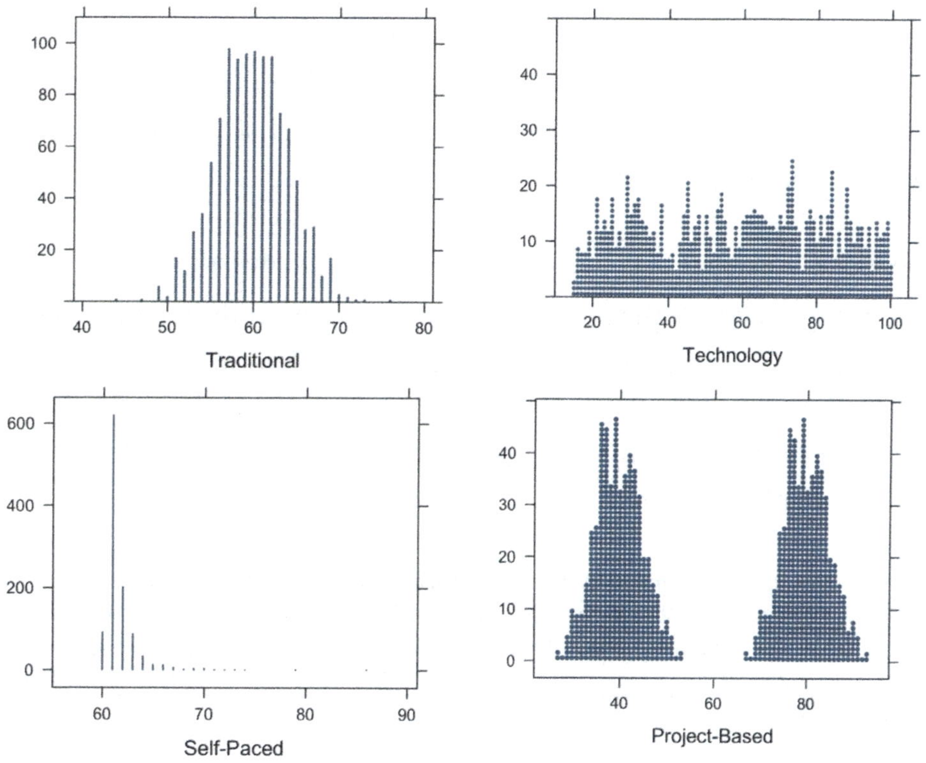

Population distributions of test scores for each of the four pedagogies

To characterize the dotplot representations of the population distributions by distribution shape, we describe the traditional pedagogy math test scores distribution as symmetric and single-mound shaped, the technology pedagogy math test scores as somewhat uniform, the self-paced as skewed to the right, and the project-based as bimodal, which indicates that the distribution has two modal clusters, as indicated by the two peaks. The population means and the population standard deviations for each of these distributions are given in the following table:

Pedagogy	Mean	Standard Deviation
Traditional	59.83	4.17
Technology	58.18	24.01
Self-Paced	61.65	1.77
Project-Based	59.63	20.62

For each of these populations, we are going to simulate taking 1000 random samples, compute the means for each sample, then use these sample means to create a dotplot representation of the sampling distribution, which will then allow us to visualize the approximate sampling distribution of the sample mean. We are going to repeat this process for different fixed sample sizes. First, we will do this for samples of size 10. Next we will use samples of size 30, and lastly we will use samples of size 50. For each population and each sample size, we will end up with a dotplot representation of the approximate sampling distribution for the sample mean. Thus, we will have a total of 12 approximate sampling distributions for the sample means.

The purpose of this exercise is to answer the following two questions:

1. How does the shape of the population distribution affect the sampling distribution?
2. How does the sample size affect the sampling distribution?

As we construct these approximate sampling distributions, we will fill in the following table to keep track of the characteristics of the sampling distributions.

Population	Size of Sample	Shape of the Approximate Sampling Distribution	Mean of the Approximate Sampling Distribution (mean of the sample means)	Standard Deviation of the Approximate Sampling Distribution (standard deviation of the sample means)
Traditional	$n = 10$ $n = 30$ $n = 50$			
Technology	$n = 10$ $n = 30$ $n = 50$			
Self-Paced	$n = 10$ $n = 30$ $n = 50$			
Project-Based	$n = 10$ $n = 30$ $n = 50$			

For the traditional teaching method, here are the three simulated approximate sampling distributions:

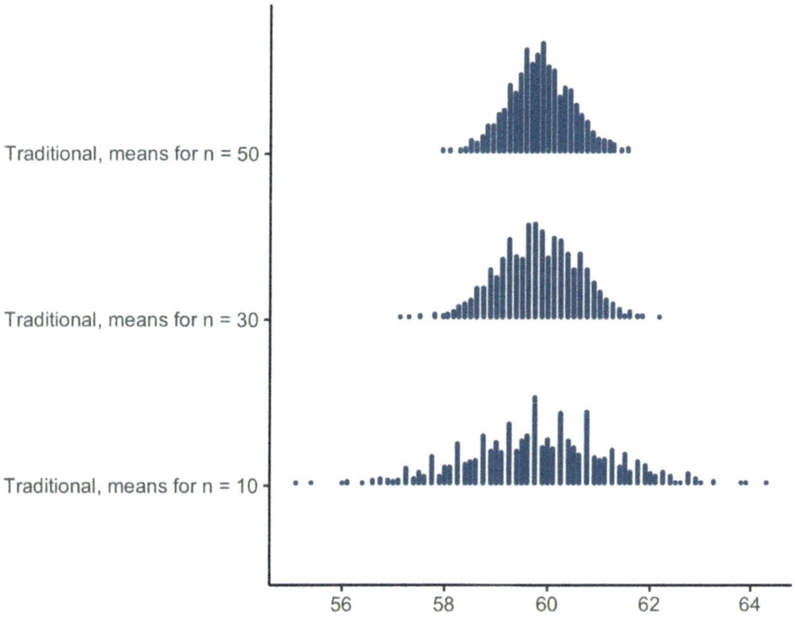

Approximate sampling distributions for the sample mean with samples of size 10, 30, and 50 for traditional pedagogy

Now that we have our three approximate sampling distributions, we can report on the shape, variability, and center of the distribution. We see that as the sample size increases, the shape of the sampling distribution tends to have less variability and is single-mounded and symmetric. The center of the sampling distributions stays about the same for all sample sizes. The descriptive statistics to fill into the table for the traditional teaching method are:

	n	Mean of the Sampling Distribution	Standard Deviation of the Sampling Distribution
n = 10	1000	59.82	1.32
n = 30	1000	59.81	0.77
n = 50	1000	59.83	0.60

We now repeat the same process with the next teaching pedagogy. For the technology-based teaching method, here are the three simulated approximate sampling distributions:

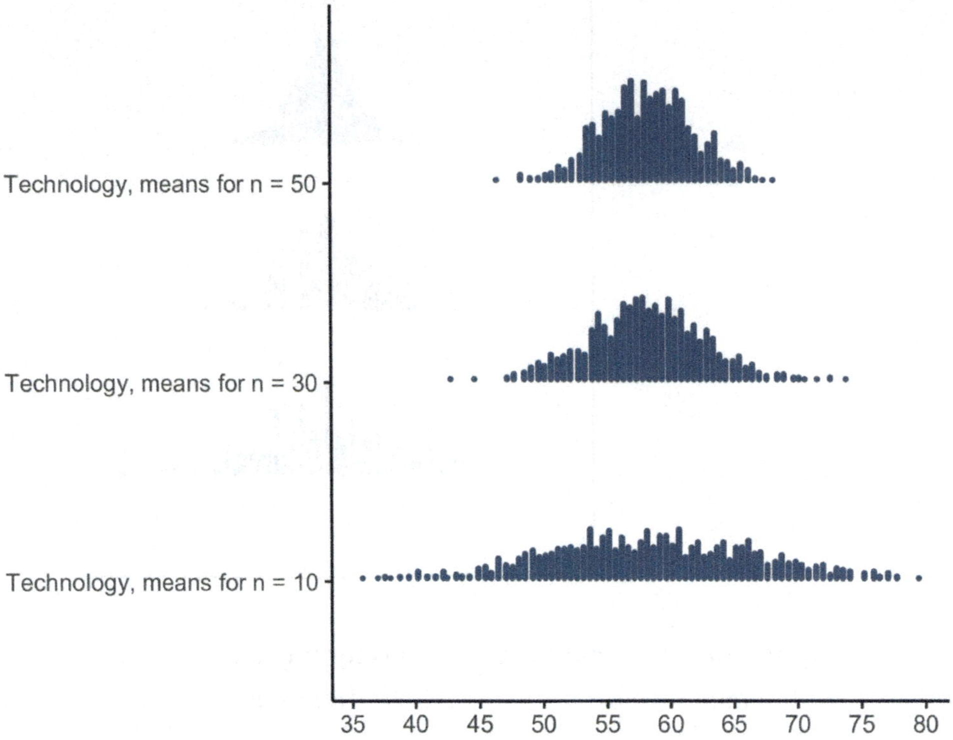

Approximate sampling distributions for the sample mean with samples of size 10, 30, and 50 for technology pedagogy

Again, we see that as the sample size increases, the shape of the distribution indicates less variability and is single-mounded and symmetric. The descriptive statistics to fill into the table for the technology teaching pedagogy are:

	n	Mean of the Sampling Distribution	Standard Deviation of the Sampling Distribution
n = 10	1000	58.35	7.83
n = 30	1000	58.13	4.42
n = 50	1000	58.20	3.47

This table also shows that the means are similar regardless of sample size, and that the standard deviation of the sampling distribution decreases as the sample size increases.

For the self-paced method, here are the three simulated approximate sampling distributions:

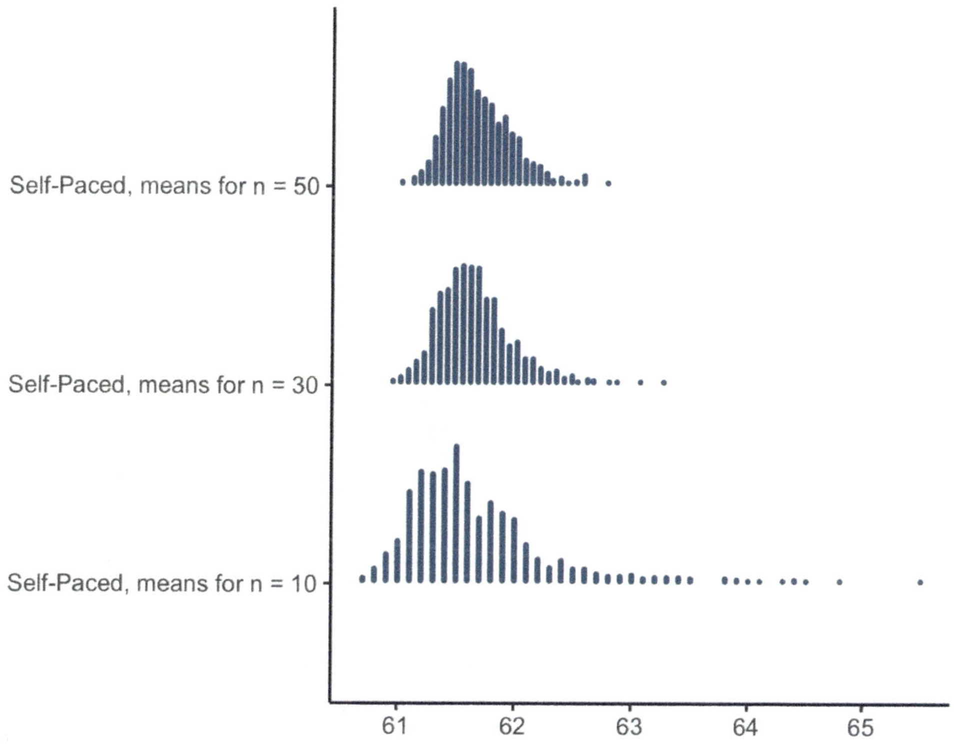

Approximate sampling distributions for the sample mean with samples of size 10, 30, and 50 for self-paced pedagogy

The self-paced approximate sampling distributions seem to all have a tail. The tail decreases as the sample size increases and the approximate sampling distribution appears to become more symmetric. The descriptive statistics to fill into the table for the self-paced teaching method are:

	n	Mean of the Sampling Distribution	Standard Deviation of the Sampling Distribution
n = 10	1000	61.64	0.57
n = 30	1000	61.64	0.30
n = 50	1000	61.66	0.27

We see that the center stays pretty much the same in the approximate sampling distributions for all of the sample sizes. Again, the variability decreases as the sample size increases, and the shape has less of a tail and becomes more symmetric about its single mound. For the project-based teaching method, here are the three simulated approximate sampling distributions:

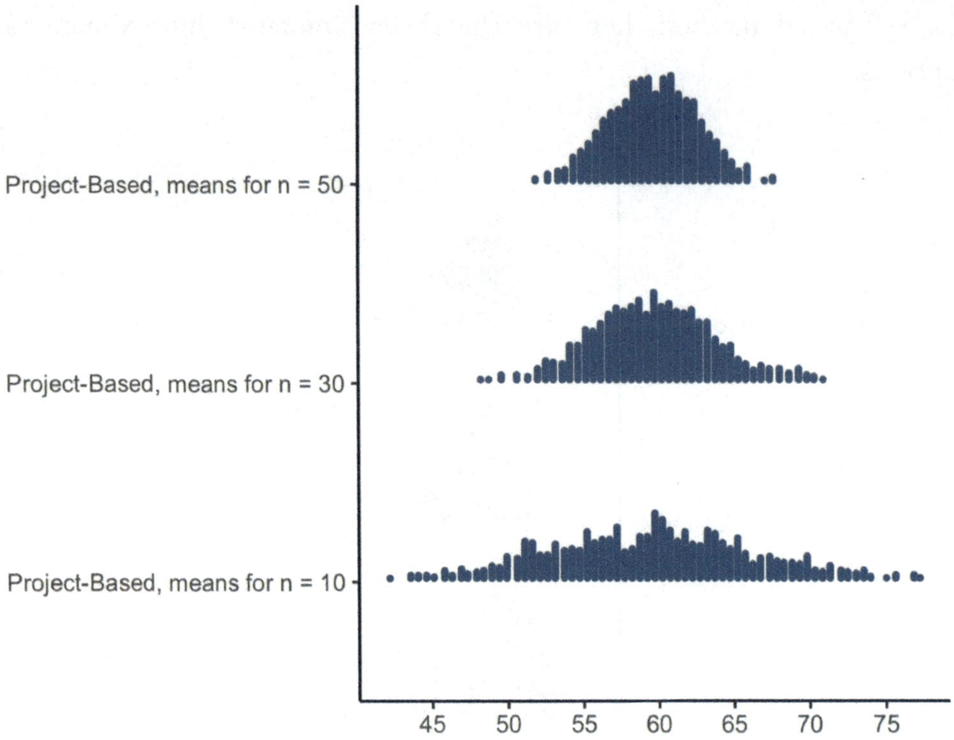

Approximate sampling distributions for the sample mean with samples of size 10, 30, and 50 for project-based pedagogy

The shape of all these distributions are approximately symmetric and single-mounded. The descriptive statistics to fill into the table for project-based teaching pedagogy are:

	n	Mean of the Sampling Distribution	Standard Deviation of the Sampling Distribution
$n = 10$	1000	59.40	6.45
$n = 30$	1000	59.64	3.86
$n = 50$	1000	59.66	2.83

We encourage teachers and students to work in groups of four, in which each person in the group chooses a pedagogy, then carries out the steps outlined previously. For each pedagogy, we want three approximate sampling distributions constructed (one for sample sizes of $n = 10$, 30, and 50) and the descriptive statistics computed. If working in groups, then each group member will have this information for their chosen pedagogy, and the investigation can continue by comparing the results that each group member found.

At this point in the investigation, we have the following four dotplots for each pedagogy:

- An approximation of the sampling distribution of sample mean x for $n = 10$
- An approximation of the sampling distribution of sample mean x for $n = 30$

- An approximation of the sampling distribution of sample mean x for $n = 50$
- The entire population distribution of 1,079 test scores for the pedagogy

Let's examine the results and answer the following questions:

- Do the sampling distributions look similar or different across the different pedagogies?
- Can the shape of a sampling distribution be predicted?
- What happens to the shape of the sampling distributions as the sample size gets larger?
- Does a larger sample size affect the mean of the sampling distribution?

The filled in table of results for our simulated approximate sampling distributions is:

Population	Size of Sample	Shape of the Approximate Sampling Distribution	Mean of Approximate Sampling Distribution (mean of the means)	Standard Deviation of Approximate Sampling Distribution (standard deviation of the sample means)
Traditional				
	n = 10	bell-shaped	59.82	1.32
	n = 30	bell-shaped	59.81	0.77
	n = 50	bell-shaped	59.83	0.60
Technology				
	n = 10	bell-shaped	58.35	7.83
	n = 30	bell-shaped	58.13	4.42
	n = 50	bell-shaped	58.20	3.47
Self-Paced				
	n = 10	skewed right	61.64	0.57
	n = 30	almost symmetric and bell-shaped	61.64	0.30
	n = 50	almost symmetric and bell-shaped	61.66	0.27
Project-Based				
	n = 10	bell-shaped	59.40	6.45
	n = 30	bell-shaped	59.64	3.86
	n = 50	bell-shaped	59.66	2.83

From the table we can see that the shape of the original population distribution has minimal impact on the shape of the sampling distribution. In the case of the skewed self-paced approximate sampling distributions, we see a right tail, but that tail diminishes as the sample size increases. The sampling distributions appear to approach a symmetric, single-mounded distribution (approximate normal distribution) as the sample size increases. This enables us to conclude that the *shape* of the sampling distribution of the sample mean can, in fact, be predicted. We expect the shape to be modeled by a normal distribution, provided that the sample size is large enough. Additionally, in all cases, the mean of the sampling distribution is approximately equal to the mean of the population (seen in column four of the table). We also note that the standard deviation of the sampling distribution, referred to as the **standard error**, decreases as the sample size increases for all of the four instructional pedagogies.

Sampling distributions simulated in this investigation are approximate. The theoretical sampling distribution would consider all possible samples from a population of a certain size. This investigation shows that it is important to understand that *regardless* of the shape of the population distribution, as the sample size increases, the shape of the sampling distribution for the mean is bell-shaped. The variability of the sampling distribution is also dependent on the sample size (the larger the sample size, the smaller the standard deviation decreases), and the mean of the sampling distribution is not dependent on the sample size. Instead, the mean of the sampling distribution, for any size sample, is expected to be equal to the population mean. The next investigation further develops these main concepts and unites them into an important theorem in statistics called the central limit theorem.

INVESTIGATION SUMMARY

The main concepts developed in the different pedagogies investigation are:

1. The mean of the sampling distribution of the sample mean does not depend on the sample size. In theory, it is always equal to the mean of the population.
2. The variability of the sampling distribution of the sample mean decreases as the sample size increases.
3. The shape of the sampling distribution of the sample mean approaches a bell-shaped symmetry as the sample size increases, regardless of the shape of the population distribution.

Up to now, we have seen how to construct sampling distributions, as well as how the population distribution does not matter in predicting the sampling distribution for the mean with large enough sample size, and we have begun making connections between sampling distributions and inference. In this activity, we will be formalizing the features of the sampling distribution of the mean with the **central limit theorem**.

Investigation 3C.3: The Central Limit Theorem[8]

Goals of this investigation: Derive the central limit theorem.

According to the www.pennies.org article on "A Brief History of the U.S. Cent," the first one-cent coin was created in 1787 and was made of pure copper. From 1787 to 2000, approximately 300 billion one-cent coins were created. In the name of statistics, students at a high school were instructed to bring in any pennies they had lying around as change in their house. A total of 499 pennies were collected from students. For every penny that was brought in, the year and age of the penny was recorded. The age of the penny was computed by subtracting the year the penny was made from 2014. The data were collected in Pennies.csv.

For the purpose of this investigation, you are going to suppose that the 499 pennies make up the entire population of pennies that are currently in circulation. During the investigation, we will observe how the shape, mean, and standard deviation of the sampling distribution of the mean age of the pennies differ from those of the distribution for the population of the pennies, and how the shape and standard deviation depend on the sample size. This activity will lead us to discover the central limit theorem.

Let's begin by visualizing the population distribution of penny ages. The population distribution for the penny ages is:

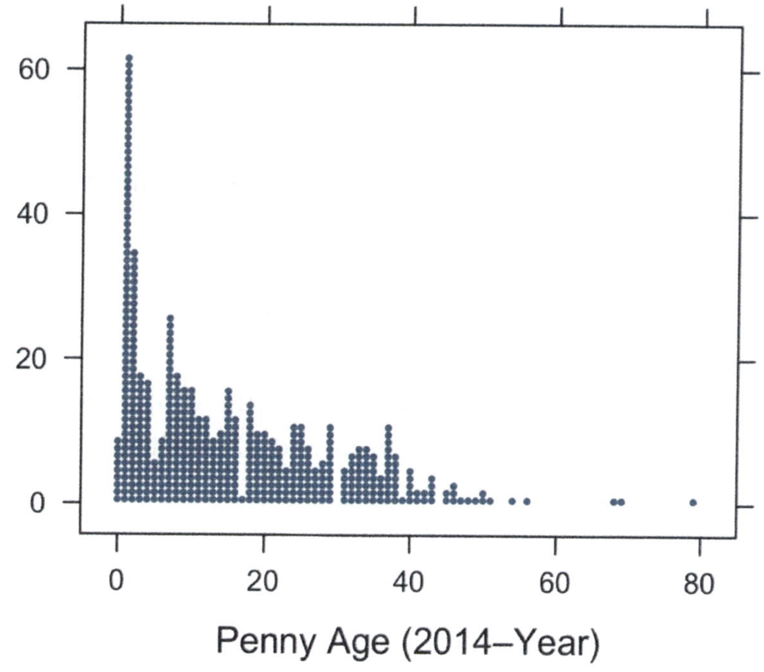

8 This pennies activity was adapted from Scheaffer, R., Gnanadesikan, M., Watkins, A., and J. Witmer. 1996. *Activity-Based Statistics*. New York: Springer.

The distribution is right skewed, which makes sense because it would be more common to have newer pennies in circulation rather than older pennies. The population mean and population standard deviation are:

	Population Size	Mean	Standard Deviation
Penny Age	499	15.72	13.8

To compare the approximate sampling distribution for the sample mean and the population distribution, we will construct approximate sampling distributions for the sample mean age of the pennies for samples sizes of 10, 25, and 50. To do this, simulate 1000 random samples for each sample size. These distributions were created using the methods described in this investigation and collected in Pennies.csv.

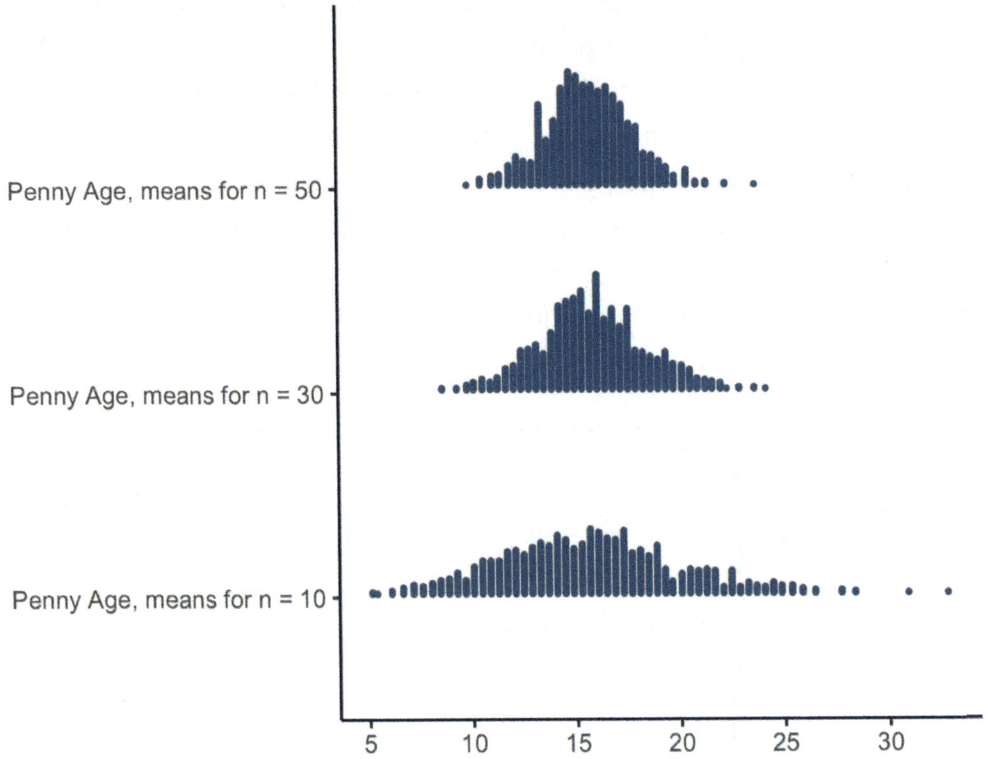

Approximate sampling distributions for the sample mean age of pennies for samples of size 10, 25, and 50 from the population

The mean and standard deviation for each of these approximate sampling distributions are:

	n	Mean of the Sampling Distribution	Standard Deviation of the Sampling Distribution
$n = 10$	1000	15.60	4.26
$n = 30$	1000	15.81	2.52
$n = 50$	1000	15.71	2.00

As we saw in the previous investigation, the mean of the sampling distribution stays about the same for each sample size, the variability (standard deviation of the sampling distribution) decreases as the sample size increases, and as the sample size increases, the shape of the distribution is approximately bell-shaped. Let's try to be even more precise about these statements.

The mean of the sampling distribution is equal to the population mean regardless of the sample size.

This finding may not be extremely obvious in these investigations because we are constructing approximate distributions. If we took *all* possible samples of the same size and constructed the entire sampling distribution (not just an approximate one), we would see this result exactly. However, even when working with approximate sampling distributions, we see that the means are consistently close to 15.6–15.7.

Our next result is the following:

There is less variability in the sampling distribution as the sample size gets larger. The standard deviation of the sampling distribution gets smaller and smaller as the sample size increases.

The standard deviation of a sampling distribution is called the **standard error** of the sample mean (or of the statistic). We see that the standard error decreases as the sample size increases directly from the previous table, but we can be more precise and try to find a formula that models how it decreases. To find a formula for the standard error, let's begin by making a graph with the sample size plotted on the horizontal axis and the standard deviations of the sampling distributions for the different size samples plotted on the vertical axis. In looking at the plot, what types of functions could possibly model the relationship (e.g., linear, quadratic, power, exponential, etc.)?

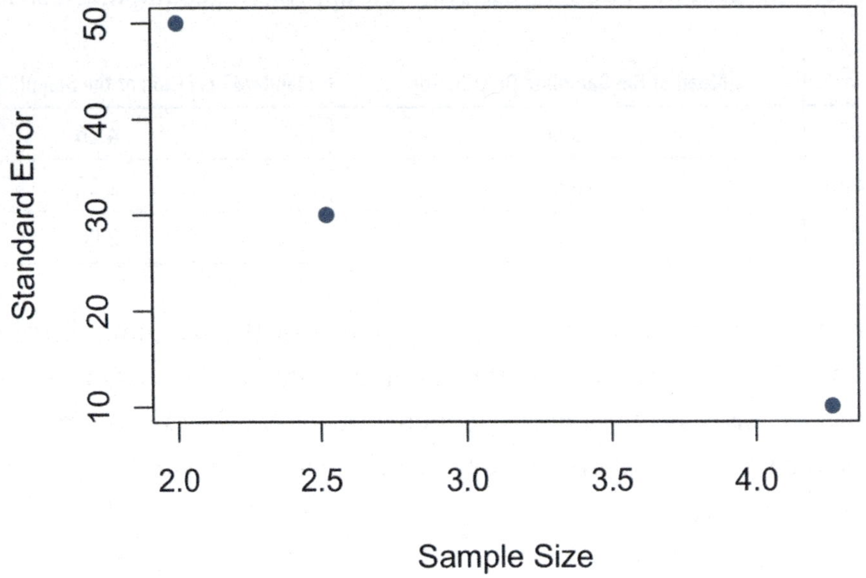

Recall that in prior units, we explored linear relationships by visualizing a piece of spaghetti or straight edge to find a linear equation. The relationship does not appear to be linear because no matter how you try to line a piece of spaghetti between the three points, a linear function does not appear to be the best fit. However, it does seem like the relationship could be modeled using a curve of some type. Using an online curve fitter, such as www.mycurvefit.com/, we can try different types of curves to fit the points. In this case, a power function provides the best fit to the graph.

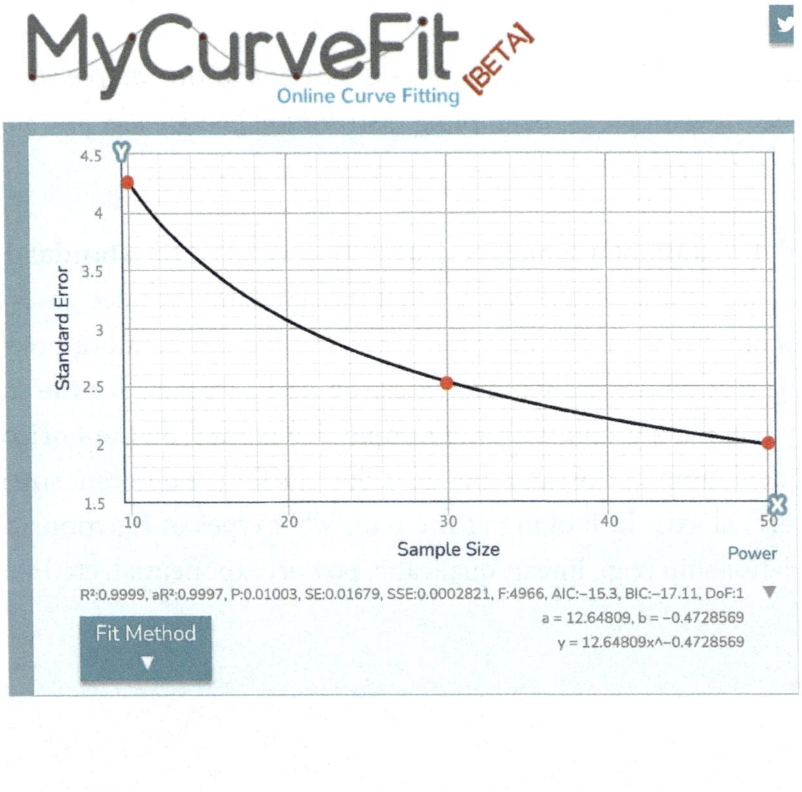

On the bottom corner of the output, we see the equation is: $y = 12.65\,x^{-0.473}$

Because the exponent –0.473 is approximately $-1/2$, we can move the x into the denominator and rewrite the function as: $y = \frac{12.65}{\sqrt{x}}$

But x is the sample size. Therefore, this function is the following: $y = \frac{12.65}{\sqrt{\text{sample size}}}$

Notice that the 12.65 is very close to the standard deviation of the population. If we had the entire sampling distribution, the numerator of the standard error would be exactly equal to the standard deviation of the population. So not only do we know that the standard error decreases as the sample size increase, but we precisely know:

$$\text{standard error} = \frac{\text{standard deviation of the population}}{\sqrt{\text{sample size}}}$$

Lastly, we can visualize the following:

As the sample size increases, the shape of the sampling distribution is bell-shaped.

Now we can put all of these statements together and formally state the central limit theorem.

The Central Limit Theorem (CLT) for the Sample Mean:

The sampling distribution of the sample mean is centered at μ; the mean of the population has a standard error $SE = \frac{\sigma}{\sqrt{n}}$, and approaches the normal distribution as the sample size increases.

In other words, even with having access to only one sample, we can use the sampling distribution to help us state whether certain values of a statistic are surprising or plausible. This is because the CLT gives us guidance on the distribution of the sample statistic. For prior investigations, we had to simulate sampling distributions for sample means, but now, because of the CLT, we can predict how the sample mean behaves. In turn, the CLT allows us to make conclusions about the population, which was our ultimate goal as described in the inference loop depicted in the inference process diagram at the beginning of this unit.

It is important to be able to describe the CLT in your own words and recognize the value of the CLT for inference.

INVESTIGATION SUMMARY

The main concept developed in the central limit theorem investigation is:

The central limit theorem states that as n, the size of the sample, increases, the standard deviation of the sampling distribution of the sample mean decreases according to the formula $\frac{\sigma}{\sqrt{n}}$, while the mean of the theoretical sampling distribution is always equal to the mean of the population. As the size of the sample increases, the shape of the sampling distribution of the sample mean can be approximated by a bell-shaped, symmetric curve.

Knowing the behavior of a sampling distribution for a statistic allows us to connect sample information to the population through inference. The result of the central limit theorem is extremely powerful in statistics because it is the crux of inference. It should be noted that the concept of a sampling distribution is challenging for people to understand. Using simulations can help develop this knowledge, but simulations can also lead people astray, because a person may think they need to always have multiple samples in order to perform inference. Being aware of this potential pitfall can help teachers phrase the ideas of this unit carefully in order to not lead students astray.

Because of the importance of the central limit theorem and this unit, a summary of the key ideas presented in the unit are included here. They are:

1. The sampling distribution is a distribution that describes how a statistic varies for repeated samples of the same size sampled from the same population.
2. The sampling distribution shows which values of a statistic are common and which are surprising.
3. The mean of the sampling distribution of the sample mean does not depend on the sample size. The mean of the theoretical sampling distribution of the sample mean equals the mean of the population.
4. The variability of the sampling distribution of the sample mean decreases as the sample size increases according to the formula $\sigma/\sqrt{n}$, where n is the sample size.
5. The shape of the sampling distribution of the sample mean looks more and more normal as the sample size increases.

While these investigations focused on the sample mean, we also obtain similar results for the sample proportion. For example, suppose we are interested in the population proportion corresponding to the proportion of people who thought favorably about the Every Student Succeeds Act (ESSA), or the proportion of students who are eating breakfast in the morning before going to school, and we randomly sample people from a population to survey them about these issues. Different samples of people will lead to different sample proportions of people who favor ESSA or eat breakfast in the morning. Therefore, the value of the sample proportion varies from sample to sample. Similar to the mean, the sampling distribution for the sample proportion also follows a normal distribution as sample sizes increase! Thus, the central limit theorem for the sample proportion can be stated as follows:

The Central Limit Theorem for the Sample Proportion:

The sampling distribution of the sample proportion is centered at p, the proportion of the population, and has a standard error $SE = \sqrt{\frac{p(1-p)}{n}}$, which decreases as the sample size increases. The shape of the sampling distribution is approximately normal for a large enough sample size n.

Follow-Up Questions

1. **Plausible/Common or Surprising/Unusual?**

Principal Brown is interested in understanding how her middle-school students compare with middle-school students nationally.

One thousand middle schools across the United States were randomly selected by the federal government to answer a survey about student backgrounds and behaviors. The middle schools were sampled from medium-size suburbs in middle- to upper-middle-class areas. The middle schools all had approximately 600 students and were considered midsize. Some of the questions on the survey were:

a. Do you participate in school-sponsored after-school activities?
b. Do you have any siblings at the middle-school or high-school grade level?
c. Did you eat breakfast this morning?
d. Did you go to bed before 11 p.m. last night?
e. Do you have a computer with internet access at home?

For each of the 1000 middle schools, we can pretend that the Department of Education printed the following graphical displays that represent the sampling distribution of the sample proportion of students replying "yes" for each survey question.

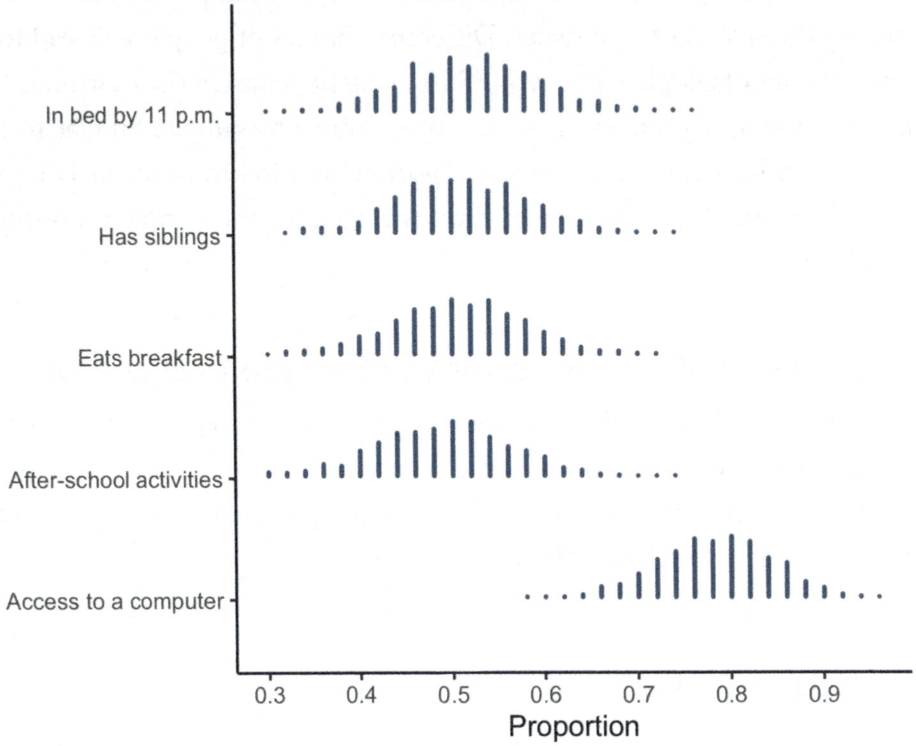

Principal Brown's school has 600 students, which is the same size as the schools sampled by the Department of Education. She decides to ask her student body the same survey questions to determine how her school compares with others nationally. Her results are displayed in School_Data.csv. Using the data set, compute the percentage of students that said "yes" for each question at Brown's school.

a. Looking at the approximate sampling distribution provided by the Department of Education, locate Brown's summary statistic on the distribution. Mark it.

b. Decide whether Brown's school's summary statistic is common or unusual. Explain why and how you made this decision.

2. Are all sampling distributions normal?

The investigations in this unit derived the central limit theorem for the sample mean. We saw that regardless of the distribution of the population from which samples are drawn, the sampling distribution of the sample mean is approximately normal. We also noted that the sample proportion behaves similarly to the sample mean. In this problem, we will examine whether this pattern exists for other statistics beyond the mean. We will investigate the answer to the following question:

Does the sampling distribution for the sample maximum behave in a similar way to the sampling distribution of the mean and proportion?

To answer the question, use the Pennies.csv data. Simulate three sampling distributions for the sample maximum (instead of computing the mean for each sample, you will compute the maximum for each sample). Take random samples of 10 pennies, 30 pennies, and 50 pennies, each 1000 times. For each sample, compute the maximum. What do you observe about the mean, standard error, and shape of each sampling distribution? Did the sampling distribution for the maximum behave in a similar way to the sampling distribution of the mean? Why do you think this is?

The previous investigations developed an approximate sampling distribution of a sample statistic from repeated sampling of same-size samples from a population. However, in reality, we do not in fact have access to repeated samples from a population, but instead access to only one sample. Using the one sample, can we get an approximate sampling distribution to give us an idea of which values of the sample statistic might be plausible and likely or surprising and unlikely? The answer is yes. There is an alternative method for obtaining insights about the sampling distribution for *any* sample statistic called the bootstrap. Note that the sampling distributions for the sample mean and the sample proportions are known but others are not.

The bootstrap method essentially takes the one random sample and treats it as a population. From this "population," we repeatedly sample *with replacement* samples of the same size. For each sample, we proceed in the same manner as the traditional methods described previously. For each sample, we compute the sample statistics and thus we can look at the Bootstrap approximate sampling distribution. The next investigation works through this idea.

Investigation 3C.4: Pennies Continued

Goals of this investigation: Introduce the bootstrap.

We aim to investigate the following question:

What are plausible values for the average age of pennies in circulation?

A total of 499 pennies were collected from students in a high-school class. For every penny that was brought in, the year and age of the penny was recorded. The age of the penny was computed by subtracting the year the penny was made from 2014. The data were collected in Pennies.csv.

For the purpose of this investigation, we are going to pretend that the 499 pennies make up the entire population of pennies that are currently in circulation. For this sample of pennies, we find that the average age of a penny is 15.7 years old. The sample distribution of the 499 pennies is pictured in the following dotplot:

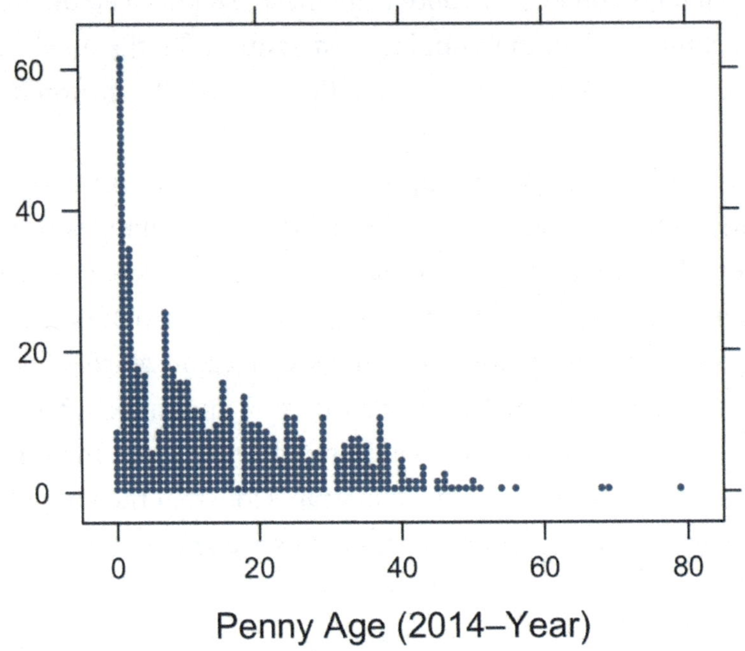

To get a sense of the plausible average age, we have to understand if getting a mean age of 15.7 years is something that is common or is surprising. To do this, we can simulate more sample means, say 1000, and see if many come out to be near 15.7. Because we do not have access to the entire population of pennies in circulation, we are going to treat this sample of 499 pennies as our population, and we are going to sample directly from it.

The key about the bootstrap is to sample with replacement. This process of sampling is such that for every penny age sampled from the 499 pennies, that penny age is put back in the "population" before the next penny age is sampled. This way a penny could be repeatedly sampled. We carry out this random sampling with replacement process 1000 times:

```
replicate(1000, mean(sample(pennies$age, size = 499, replace = TRUE)))
```

We set the sample size at 499, mimicking the repeated sample of the same size we did in the traditional methods, and simulate random samples with replacement. For each sample, we find the mean age of the pennies in the sample and store those sample means in a new column.

To visualize these sample means, we can create a dotplot (just like we did in the previous investigations) that represents the bootstrap sampling distribution:

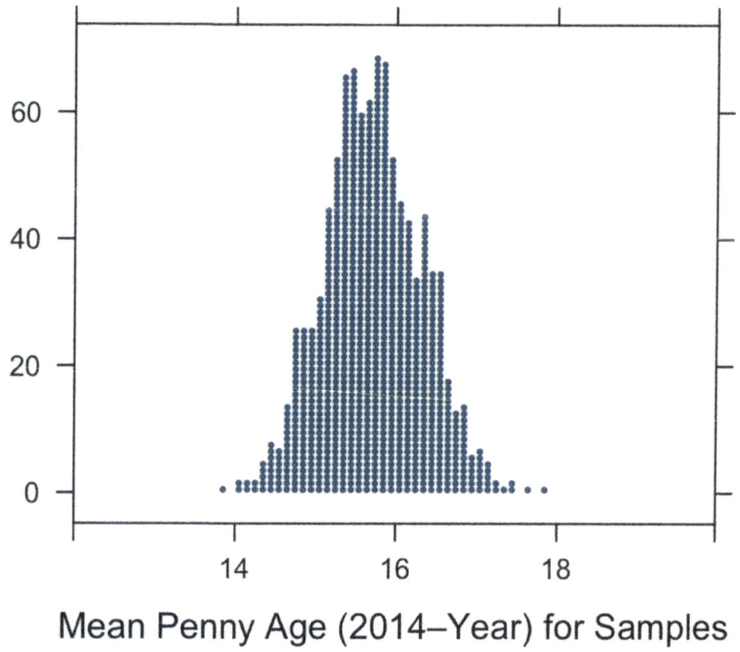

Similarly to the previous investigations, we can also compute the mean and the standard error of these sample means:

	n	Mean	Standard Deviation
Mean of the sSample Means	1000	15.76	0.62

From this, we can see that our original sample statistic value of an average age of 15.7 is in fact very plausible. This compares well with the repeated sampling information we obtained in the prior investigation, in which we found that for sample size $n = 50$, our approximate sampling distribution had a mean of 15.71 and a standard deviation of 2.00.

The bootstrap method offers another approach to finding an approximate sampling distribution in a situation where we do not have access to the entire population. It is important to note that in real-life situations, we never have access to the entire population if we are looking only at sample information. If we had information about the entire population, we would merely use it and not bother with sampling. The bootstrap method thus allows us to investigate the behavior of any sample statistic in a real setting. This is especially important for the sampling distribution of a sample statistic that cannot be modeled by the normal curve, such as the maximum value in a random sample.

At this point, we can answer the investigative question "What are plausible values for the average age of pennies in circulation?" using a confidence interval. Because the sampling distribution created with the bootstrap looks approximately normal, then we apply the empirical rule. We know that approximately 95% of the sample means are within two standard deviations (0.62) of the mean of the sampling distribution (15.76).

Thus, we can consider forming a 95% confidence interval that shows plausible values for the average age of pennies in circulation by:

$$15.72 \pm 2(0.61) = (14.5, 16.94)$$

Therefore, at the 95% confidence level, we give the range of plausible values for the average age of pennies in circulation to be (14.5, 16.94).

INVESTIGATION SUMMARY

The main concepts developed in the pennies continued investigation are:

1. The bootstrap provides an alternative approach to understanding the sampling distribution of a sample statistic that is very effective in real-life scenarios when the population is unknown and when the behavior of the sampling distribution of a statistic is unknown.
2. The bootstrap treats the original sample as the population. It necessitates the use of technology to sample with replacement from the original sample.
3. Using the approximate sampling distribution found through the bootstrap, one can estimate a range of plausible values for the population parameter of interest.

The CLT is extremely important in inferential statistics. While in the previous investigations we derived the CLT by taking repeated samples, we in fact do not do this in real life. As noted in the bootstrap investigation, we have access to only one sample. The CLT is powerful because we know information about how the sample mean or the sample proportion behaves without needing to know anything about the population. Before technology was available to use a method such as the bootstrap, the CLT provided the backbone for all inference. As technology has become more powerful, the bootstrap offers an alternative to the theory of the CLT. Through simulations, the bootstrap method can give you an idea of the sampling distribution of ***any*** statistics using sampling with replacement, as described in the investigation. While the CLT applies only to the mean and the proportion, the bootstrap approach can be used in situations where you are interested in other sample statistics.

While the bootstrap offers a nice alternative to the theoretical method of the CLT, at this time, the bootstrap is not as widely taught in education. When interested in the sample mean or sample proportion, the theory of the CLT is still applied.

References

Bargagliotti, A., Franklin, C., Arnold, P., Gould, R., Johnson, S., Perez, L., and D.Spangler. 2020. *Pre-K–12 Guidelines for Assessment and Instruction in Statistics Education II* (*GAISE II*). American Statistical Association and National Council of Teachers of Mathematics.

Scheaffer, R., Gnanadesikan, M., Watkins, A., and J. Witmer. 1996. *Activity-Based Statistics*. New York: Springer.

Watkins, A.E., Bargagliotti, A. & Franklin, C., (2014). Simulation of the sampling distribution of the mean can mislead. Journal of Statistics Education, 22(3).

Final Summary

The concepts and materials presented throughout this book are meant to address teachers' and students' learning of statistics and data science up until basic inference and more complex algorithms and computing. The concepts in this book are foundational to working with data. Further additional information and examples can be found in the *GAISE II* report, as well as in a resource guide maintained at www.nctm.org/gaise. Data sets used in this book can be downloaded at https://bit.ly/Statistics-DataScience-for-Teachers. Other important resources can be found under the Education tab at http://www.amstat.org.

Acknowledgments

A special thank you to Pip Arnold, Stephanie Casey, Rob Gould, Shonda Kuiper, Gary Kader, Steven Miller, James Perrett, Josh Tabor, Doug Tyson, and Ann Watkins for reviewing the book.

The authors would like to sincerely thank Terri Johnson for all of her help with the graphics and Caroline Coppel for editing. We also appreciate the design and layout work of Shirley E.M. Raybuck, and her artwork on the cover.

A special thank you to India Dastic, David De LaTorre, Efrain Estrada, Anna Gralnik, Jessica Quin, Rosa Pastor, Alvaro Pineda, Kim Price, and other key teachers who read, designed lesson plans, and made suggestions for improvement over the course of several years by testing the investigations in this book. Also, the authors would like to thank the countless teachers whom we have had conversations with about the topics and specific investigations presented in this book as well as all of the students who have piloted many of the lessons and contributed their work as exampleS in this book.

The authors extend a sincere thank you to the ASA/NCTM Joint Committee for always being forward thinking and funding the production process of *GAISE II* and this book.

A special thank you to Donna LaLonde and Rebecca Nichols from the ASA and Dave Barnes and Jeff Shih from the NCTM for their support throughout the process.

Anna would like to dedicate this book to her kids, Siena (11) and Luca (8), whose willingness to test knowingly and sometimes unknowingly many of the investigations in the book led to many improvements. She would also like to dedicate this to her husband Joey whose love and support kept her steady to complete this book project.

Chris would like to dedicate this book to Anna Bargagliotti, whose vision and dedication led to the writing of this book. Anna has been steadfast in her commitment to K–12 statistics education. I am thankful that Anna is my special colleague and friend.